Mobile Crowdsensing

Mobile Crowdsensing

Cristian Borcea • Manoop Talasila
Reza Curtmola

CRC Press
Taylor & Francis Group
Boca Raton London New York

CRC Press is an imprint of the
Taylor & Francis Group, an **informa** business

A CHAPMAN & HALL BOOK

CRC Press
Taylor & Francis Group
6000 Broken Sound Parkway NW, Suite 300
Boca Raton, FL 33487-2742

First issued in paperback 2020

© 2017 by Taylor & Francis Group, LLC
CRC Press is an imprint of Taylor & Francis Group, an Informa business

No claim to original U.S. Government works

ISBN-13: 978-1-4987-3844-6 (hbk)
ISBN-13: 978-0-367-65830-4 (pbk)

Visit the Taylor & Francis Web site at
http://www.taylorandfrancis.com

and the CRC Press Web site at
http://www.crcpress.com

To my wife, Mina, for her support during the writing of this book.
Cristian Borcea

This book is dedicated to my lovely daughter, Ishika Talasila, who
distracted me a lot while writing this book.
Manoop Talasila

To my family, which means everything to me.
Reza Curtmola

Contents

Preface

Mobile crowdsensing is a technology that allows large scale, cost-effective sensing of the physical world. In mobile crowdsensing, mobile personal mobile devices such as smartphones or smart watches collect data related to environment, transportation, healthcare, safety, and so on. In addition, crowdsourcing techniques are used to manage the entire mobile sensing ecosystem. Together with static sensor networks and the Internet of Things, mobile crowdsensing has the potential to make the vision of ubiquitous sensing a reality.

Mobile crowdsensing has now matured enough, with well-defined advanced systems and protocols, and this knowledge has to be made available as a book for broader audiences to understand this new technology that can change the world. The information is currently published in scientific journals and conferences, but the time has come to have a comprehensive book on this topic.

This book provides a one-stop shop for all relevant issues pertaining to mobile crowdsensing, including motivation, applications, systems and platforms, incentive mechanisms, reliability, security, and privacy, which would otherwise be available only as a disparate collection of heterogeneous sources. Its description of the design and implementation of mobile crowdsensing platforms can be of great interest for readers working in research and development to quickly develop and deploy their systems.

The book also identifies potential issues in building mobile crowdsensing applications to ensure their usability in real life. For example, it discusses details of sensing data reliability protocols, such as location authentication. Furthermore, it provides the latest incentive concepts in crowdsensing, such as mobile sensing games. Finally, it presents future directions in mobile crowdsensing and open problems that have to be addressed.

This book can serve as a useful reference for graduate students, professors, and mobile data scientists. Readers working in industry or research labs focusing on Smart City projects as well as the managers of Smart Cities will also benefit from the concepts and technology covered in this book.

We would like to acknowledge the researchers whose work is referenced in the book to broaden the scope of mobile crowdsensing. We would also like to thank our families and friends for their constant encouragement, patience, and understanding throughout this project. Finally, we thank the readers who might soon participate in mobile crowdsensing activities.

Cristian Borcea, Manoop Talasila, and Reza Curtmola

List of Figures

List of Tables

1

Introduction

Mobile crowdsensing is a technology that allows large-scale, cost-effective sensing of the physical world. In mobile crowdsensing, people going about their daily business become "sensors." Their mobile personal mobile devices such as smartphones or smart watches come equipped with a variety of sensors that can be leveraged to collect data related to environment, transportation, healthcare, safety and so on. This people-centric sensing is enabled by crowdsourcing techniques to create a large-scale, collective sensing environment. Together with static sensor networks and Internet of Things devices, mobile crowdsensing has the potential to make the vision of ubiquitous sensing a reality.

This book presents the first extensive coverage of mobile crowdsensing, with examples and insights drawn from the authors' extensive research on this topic as well as from the research and development of a growing community of researchers and practitioners working in this emerging field. To make sure that the book remains grounded in reality, we have decided to provide a multitude of application, system, and protocol examples throughout the text. Our insights are always drawn from practical scenarios and real-life experience.

To introduce the readers to mobile crowdsensing, this chapter starts with a short history of sensing, with a focus on wireless and mobile sensing. We then discuss how ideas from two distinct areas, mobile people-centric sensing and crowdsourcing, have led to the concept of mobile crowdsensing. The chapter ends with an overview of the rest of the book.

1.1 Evolution of Sensing

Sensors have been used for a long time in industries such as car manufacturing and electric power generation. These sensors are highly specialized, require fine tuning and calibration, and are wired to other control equipment. Naturally, they are costly and cannot be used efficiently in a wide spectrum of application domains, such as environment monitoring, agriculture, and road transportation.

With advances in computer power and wireless communications, wireless sensors were introduced in the late 1990s to replace and expand the use of wired sensing. Although they are called sensors, these devices are small computers with sensing and networking capabilities. In addition, they are battery powered. Two main factors have driven the research and development of wireless sensors: cost and scale. Since they do not need wires for power and control, they can be deployed across large areas and, thus, lead to ubiquitous sensing. Instead of having complex and costly control equipment to manage the sensors and collect their data, wireless sensors contain intelligence that drives their functionality and self-organize into networks that achieve the required tasks in a distributed fashion.

Wireless sensors and sensor networks have been deployed by many organizations for monitoring purposes. For example, they have been used to monitor the structural integrity of buildings or bridges, the energy consumption in buildings, the parking spaces in a cities, or the behavior of plants and animals in different environment conditions. However, after about a decade of using wireless sensors networks, it became clear that the vision of ubiquitous sensing is not going to be achieved by this technology alone.

Three problems have precluded wireless from widespread deployment: battery power limitations of sensors, the close-loop and application-specific nature of most network deployments, and the cost. Unlike computing and networking technologies, which have improved exponentially over the last half century, the battery capacity has increased only linearly. Therefore, the lifetime of wireless sensor networks is limited by the available battery power. This makes heavy-duty sensor networks impractical in real life because changing the batteries of sensors deployed in areas that are not easily accessible is difficult and costly.

The alternative to replacing the sensor batteries would be to just replace the sensors, but this is also costly because the cost per sensor did not become as low as predicted. The reason is mostly due to economics: Without mass adoption, the cost per sensor cannot decrease substantially.

The cost per wireless sensor could be reduced if each sensor platform would have several types of sensors and multiple organizations would share a sensor network for different applications. The practice contradicts this approach, as most deployments are application specific, contain only one type of sensor, and belong to one organization. Security and privacy are two major issues that make organizations deploy close networks. The cost of providing strong security and privacy guarantees is reflected in more battery power consumption, and subsequently, lower network lifetime. Furthermore, sharing the networks for multiple applications also results in more battery consumption.

Mobile sensing has appeared in the last 10 years as a complementary solution to the static wireless sensor networks. Mobile sensors such as smartphones, smart watches, and vehicular systems represent a new type of geographically distributed sensing infrastructure that enables mobile people-centric sensing. These mobile sensing devices can be used to enable a broad spectrum of applications, ranging from monitoring pollution or traffic in cities to epidemic disease monitoring or real-time reporting from disaster situations.

This new type of sensing can be a scalable and cost-effective alternative to deploying static wireless sensor networks for dense sensing coverage across large areas. Compared to the tiny, energy-constrained sensors of static sensor networks, smartphones and vehicular systems can support more complex computations, have significant memory and storage, and offer direct access to the Internet.

Once we have this mobile sensing infrastructure, the question is: How can we leverage it into useful applications? Fortunately, another technology was developed in parallel with mobile sensing: crowdsourcing. This technology uses the collective intelligence of the crowds to solve problems that are difficult to solve using computers or are expensive to solve using traditional means such as paying experts. Crowdsourcing provides a number of techniques to recruit and incentivize a large number of participants, who can then work on subtasks of a global task. This is exactly what mobile sensing needed to achieve its full potential.

1.2 Bridging the Sensing Gap with Mobile Crowdsensing

Mobile crowdsensing combines mobile sensing and crowdsourcing to create a planetary-scale, people-centric sensing environment. A number of mobile crowdsensing applications have already proved the potential of this technology. For example, Waze [32] and other traffic applications collect data from drivers' smartphones to create an accurate view of the traffic, and then compute routes that allow drivers to avoid congested spots. Similarly, many apps enable citizens to submit reports during disasters such as earthquakes and hurricanes. The data is then shared with the authorities and the public. For instance, the Federal Emergency Management Agency (FEMA) in the United States provides such an app [7].

To make many more such applications usable in real life, we need to overcome concerns expressed by potential clients (i.e., people or organizations that need sensed data) regarding the scale at which these applications can work, as well as their dependability. For example, two major questions are: Can I find enough people willing to provide data to satisfy the spatial and temporal constraints of the application? Can I trust that the sensed data is reliable? The answers lie in part with the potential participants who provide data, but they have their own concerns regarding incentives and privacy: What do I get if I consume my time and resources to participate in these applications? How can I control when, where, and what data is collected from my mobile device? What guarantees do I have that my privacy is not violated? One question that subsumes all these concerns is: Can a mobile crowdsensing system be designed to achieve a good balance between usability, privacy, performance, and reliability?

While traditional crowdsourcing platforms such as Amazon's Mechanical Turk (MTurk) [3] may seem suitable to address some of these concerns (e.g., getting access to many participants that provide data), they cannot be used directly, for a variety of reasons. First, they are not designed to leverage the sensors on mobile devices. Second, mechanisms such as task replication to ensure data reliability may not be applicable due to the context-sensitivity of the collected data. Third, traditional crowdsourcing tasks and mobile crowdsensing tasks are fundamentally different in nature. As a result, traditional crowdsourcing platforms simply act as "dumb" mediators between clients and participants, as the tasks submitted by clients are presented unmodified to participants, whereas a mobile crowdsensing platform requires more "intelligence," as global tasks submitted by clients may need to be decomposed into elementary sensing tasks to be performed by participants. This implies that, depending on participant availability, global tasks may be fully or partially completed, and it impacts the mechanism used to deal with misbehaving participants. Finally, privacy concerns are more important in mobile crowdsensing, as participants carry the mobile devices, and sensing may reveal highly personal information such as location and over-the-limit car speed. This is not necessarily a concern in traditional crowdsourcing (e.g., a typical task like transcribing an audio segment into text poses no privacy concerns).

Therefore, new technologies are developed to support real-world, large-scale, dependable, and privacy-abiding mobile crowdsensing sensing.

1.3   Organization of the Book

The rest of the book is organized as follows. We start with a description of the two main technologies that led to mobile crowdsensing, specifically mobile sensing (Chapter 2) and crowdsourcing (Chapter 3). We then present the main ideas of mobile crowdsensing (Chapter 4), illustrate several crowdsensing systems and platforms (Chapter 5) for crowdsensing, and discuss general design guidelines for crowdsensing (Chapter 6). The book continues with the three main challenges of crowdsensing systems, incentives for participants (Chapter 7), security (Chapter 8), and privacy (Chapter 9). The book ends with conclusions and a look at future directions for crowdsensing (Chapter 10).

In the following, we present a brief overview of the book chapters.

Mobile Sensing This chapter presents background information on mobile sensing and investigates the crucial components of the mobile sensing systems. It then reports on the current usage scenarios and the hurdles facing mobile sensing.

Crowdsourcing This chapter analyzes crowdsourcing, starting from its modest beginnings in the mid-2000s to its many usage scenarios encountered today. We illustrate the main crowdsourcing application domains and research areas, and then identify a number of practical challenges that are also faced in crowdsensing.

What Is Mobile Crowdsensing? This chapter defines mobile crowdsensing and explains how it solves the problems of existing mobile sensing systems. It also clarifies the shortcomings of traditional crowdsourcing systems if they are to be used for crowdsensing. Furthermore, the chapter presents emerging crowdsensing application domains and classifies different types of crowdsensing applications.

Systems and Platforms This chapter illustrates in detail three crowdsensing systems, McSense, Medusa, and Vita. We look at their architectural similarities and differences, task execution models, and types of applications they support. At the end of the chapter, we briefly discuss other noteworthy crowdsensing systems and platforms.

General Design Principles and Example of Prototype This chapter synthesizes the lessons learned from the previous two chapters, and presents general design principles for crowdsensing systems in terms of both system architecture and system implementation. We then discuss a prototype implementation of the McSense system to help the reader understand a number of practical, implementation-specific aspects of crowdsensing. Finally, the chapter discusses a major factor that influences the performance of crowdsensing systems, namely resource management on the mobile devices.

Incentive Mechanism for Participants This chapter covers three types of incentive mechanisms for participants in mobile crowdsensing: social, monetary, and game-centric. We describe the experience gained from running two user studies with monetary and game-centric incentives, and then identify for which type of tasks these two incentive mechanisms work best. The chapter concludes with a discussion on the generality of our incentive mechanism insights.

Security Concerns and Solutions This chapter presents security concerns specific to mobile crowdsensing, with a focus on data reliability in the presence of malicious participants. We discuss in detail two solutions for data reliability based on location and co-location authentication. In addition, we briefly cover a few other solutions for data validation.

Privacy Issues and Solutions This chapter tackles privacy issues in mobile crowdsensing. Unlike crowdsourcing where the participants do not disclose highly sensitive personal data, crowdsensing data includes information such as user location and activity. The chapter presents privacy-preserving architectures, privacy-aware incentives, and solutions for location and context privacy.

Conclusions and Future Directions This chapter briefly summarizes the current state of mobile crowdsensing and looks toward the future. The chapter identifies challenges at the intersection of privacy, data reliability, incentives, and resource management. Crowdsensing is expected to become a mainstream technology in smart cities, where it will complement the Internet of Things.

2

Mobile Sensing

2.1 Introduction

Ubiquitous mobile devices such as smartphones are nowadays an integral part of people's daily lives for computing and communication. Millions of mobile apps are made available to the smartphone users through app stores such as Android Play Store and iPhone AppStore. These mobile apps leverage the cameras, microphones, GPS receivers, accelerometers, and other sensors available on the phones to sense the physical world and provide personalized alerts and guidance to the smartphone users. This chapter presents the background of mobile sensing and its applications to everyday life.

2.2 How Did We Get Here?

2.2.1 Static Wireless Sensor Networks

Wireless sensor networks (WSNs) are used to sense a given target environment, and examples of sensing applications based on WSNs are intrusion detection [160], border surveillance [83], infrastructure protection [178], and scientific exploration [146].

Such sensor network applications generally deploy mobile sensors to leverage their movement capabilities. Furthermore, the mobile sensors can create self-organized, scalable networks [42]. In the border surveillance application, sensor nodes need to be deployed evenly in the given area. This deployment ensures the complete coverage of the protected area. Manually deploying static sensors evenly in a wide area is not feasible. Hence, mobile sensors are deployed near their initial deployment locations [171] and then these mobile nodes move to their accurate positions using localization systems [91, 82]. This helps in providing uniform sensing coverage without any coverage gaps on a given area in these applications.

Sensor nodes are deployed with very limited energy in their batteries, and they require a continuous energy supply to ensure the long-term functionality of WSNs. Many research projects have proposed energy conservation schemes [35], [162] to address this problem. However, such solutions are not good enough for long-running applications.

2.2.2   The Opportunity of Mobile People-Centric Sensing

Despite their many benefits, static WSNs are expensive to deploy across large areas. Therefore, many organizations are looking for alternative solutions. In addition, WSNs have two more problems: They are traditionally application-specific and do not share data outside the organization that deployed them. People-centric mobile sensing represents a cost-effective alternative (or complementary) solution to static wireless sensor networks. This type of sensing can be done with smartphones, smart watches, wearable sensors, and vehicular embedded systems, which have multiple sensors that can perform a wide variety of useful sensing. The phones, due to their ubiquity, are the major driving force in this direction. According to a forecast for global smartphone shipments from 2010 to 2018, more than 1.8 billion phones are expected to be shipped worldwide [10]. These mobile sensing devices can be used to enable a broad spectrum of applications, ranging from monitoring pollution and traffic in cities to epidemic disease monitoring, or real-time reporting from disaster situations.

2.2.3   Mobile Sensing Background

To understand the concept of mobile sensing, let us start with the high-level components involved in it. *Mobile sensing* involves a mobile application running on the user's smartphone to access sensed data from various embedded sensors, and utilizing these data for *personal* monitoring, or for *public* use by reporting the data to central servers or the cloud. In order to perform such sensing, the smartphones require operating system support, in terms of software API, to access the sensors and functions for data aggregation, analysis, and reporting.

In *personal mobile sensing*, the mobile application only runs on the user's smartphone to provide customized alerts or assistance and does not share any sensed data to central servers. In *public sensing*, the mobile applications running on many smartphones share the collected data with central servers for analysis, such that the aggregated useful public information can be shared back with the users on a grand scale. In the public sensing model, a central administrator controls the sensing activities and the sharing of the useful data. The public sensing can be achieved either automatically or through manual user actions.

2.2.4   Crucial Components that Enable Mobile Sensing

The components playing a crucial role in mobile sensing are *mobile applications (apps)*, *sensors*, and the supporting *software APIs*, as shown in Fig-

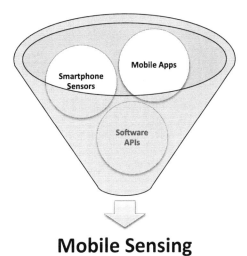

Mobile Sensing

FIGURE 2.1
Combination of components that led to mobile sensing.

ure 2.1. In first-generation smartphones, the mobile applications have been developed with the purpose of filtering web data and showing useful information that fits on the smartphone's small screens for the user's convenience. But, when the sensors started to be embedded in smartphones, the mobile applications began leveraging the sensor's data to improve the user's mobile experience. Eventually, when the smartphone operating systems started providing sensor APIs to the mobile application developers, the concept of mobile sensing took off with unlimited opportunities for sensing applications, ranging from personal monitoring to environmental and infrastructure monitoring.

2.2.4.1   Apps

The types of *mobile applications* that involve sensing are 1) personal sensing apps (such as personal monitoring and healthcare applications), 2) home sensing apps (such as HVAC or surveillance applications for smart homes), 3) city-wide sensing apps (vehicular traffic and infrastructure-monitoring applications for smart cities), 4) vehicle sensing apps (phone-to-car and car-to-phone communication sensing apps for smart cars), and 5) games (sensing for augmented reality games). Initially, the mobile application developers and the researchers have shared the common difficulty of reaching a large number of users with their sensing apps. Fortunately, companies such as Apple and Google introduced their own app stores, which created app ecosystems that made it easy for developers to publish their apps and for users to access them.

Now, the sensing applications can easily be found, downloaded, and installed by the users who take advantage of the app store ecosystems. Before

TABLE 2.1

The broad categories of sensors, radios, and other hardware available on smartphones for mobile sensing.

Motion/Position sensors	Environmental sensors	Radios	Other hardware
Accelerometer	Ambient light sensor	GPS	Microphone
Magnetometer	Barometer	Bluetooth	Camera
Gyroscope	Temperature sensor	WiFi	Camera flash
Proximity sensor	Air humidity sensor	Cellular	Touch sensor
Pedometer	Radiation sensor		Fingerprint

app stores, researchers used to perform user studies that required sensing data from people in controlled environments with small numbers of users. Today, with the success of app stores, researchers can provide their sensing apps to large numbers of users and, thus, increase the scale of their studies substantially. For instance, developers and researchers who are interested in building accurate activity recognition models can now easily collect huge amounts of sensor data from many user smartphones through their sensing apps.

2.2.4.2 Sensors

The indispensable component of mobile sensing is the *sensors*. The typical sensors available in most of the current smartphones are listed in Table 2.1. The motion sensors measure acceleration forces and rotational forces along the three axes. The position sensors, such as magnetometers, measure the physical position or direction of a device. The environmental sensors measure various environmental parameters, such as ambient air temperature and pressure, illumination, and humidity.

The radios available in smartphones are also used to sense the location, speed, and distance traveled by users. The other hardware available in smartphones for sensing are microphones and cameras for multimedia sensing, camera flashes for sensing heart-rate, touch screen sensors for activity recognition, and fingerprint sensors for security. In addition to these hardware sensors, the phones provide software-based sensors that rely on one or more hardware sensors to derive their readings (one such software-based sensor is the orientation sensor, which relies on data from the accelerometer and magnetometer to derive the phone's orientation data).

Motion/Position sensors: The most commonly available and used motion sensor is the *accelerometer*, which measures the acceleration applied to the device on all three physical axes (x, y, and z), including the force of gravity. These sensor readings are mostly used for motion detection (shake, tilt, etc.), such that the user's phone screen can be automatically re-oriented between the landscape and the portrait view. The *magnetometer* is available in most phones and measures the ambient geomagnetic field for all three physical axes (x, y, z) to detect the planet's geomagnetic north pole, which is useful for determining the direction of motion of a phone. These readings are also used by some apps for metal detection.

The *gyroscope* is another motion sensor that detects orientation with higher precision by measuring a phone's rate of rotation around each of the three physical axes (x, y, and z). These sensor readings are commonly used for rotation detection (spin, turn, etc.), which is employed, for example, in mobile games. In some phones, the orientation is also derived based on data from accelerometer and magnetometer sensors. The combination of orientation readings and location readings can help location-based applications position the user's location and direction on a map in relation to the physical world.

The *proximity sensor* measures the proximity of an object relative to the view screen of a device. This sensor is typically used to determine whether a handset is being held up to a person's ear. This simple form of context recognition using proximity sensor readings can help in saving the phone's battery consumption by turning off the phone screen when the user is on call. The *pedometer* is a sensor used for counting the number of steps that the user has taken since the last reboot while the sensor was activated. In some devices, this data is derived from the accelerometer, but the pedometer provides greater precision and is energy-efficient.

Environmental sensors: The *ambient light sensor* measures the illumination (i.e., ambient light), such that the device can adjust the display's brightness automatically, which can help in saving the phone's battery consumption. These sensor readings in combination with GPS readings and time can be used for context recognition to enhance the context-based interaction with the smartphone user. For example, the user can be greeted by the smartphone personal assistant (such as Google's Now, Apple's Siri, and Microsoft's Cortana) with a friendly message, such as "Enjoy the sunny day!!", when it detects the user is walking out from a building into a sunny open space.

The *barometer* measures the atmospheric pressure, which can be used to determine the altitude of the device above the sea level. These readings in combination with GPS, WiFi, and pedometer readings can enhance the accuracy of the user's location inside buildings. The ambient *temperature* and *humidity* sensors measure the ambient air temperature and humidity. The barometer, ambient temperature, and humidity environmental sensors are rarely found in today's phones, but they may become common in the near future. Most phones have a second temperature sensor, which is used to monitor the temperature inside the device and its battery. When the phone is overheating, it switches off to prevent damage.

The *radiation sensor* is another rare sensor which is so far available only in one smartphone [27] released in Japan. It is capable of detecting harmful radiation by measuring the current radiation level in the area.

Radios: The smartphones have GPS, Bluetooth, WiFi, and cellular radios, which are commonly used to enable location and communication features. These radios, in combination with other sensors, can enhance the user's mobile experience and can extensively contribute to the users' co-location data. Specially, the GPS location is almost always tagged to all sensing data to make the data more meaningful and context-aware.

Other hardware: Additional hardware can be used in mobile sensing, such as the microphone for sound sensing, camera for photo and video capturing (used in participatory sensing), flash light for heart rate monitoring by detecting the minute pulsations of the blood vessels inside one's finger, touch sensor for activity recognition on the screen, and fingerprint sensor, most often used as an extra layer of security (it can be used instead of a lock-screen password).

External sensors: In addition to internal smartphone sensors, sensing can be achieved through external sensors such as body-wearable sensors in the health domain to monitor patients, sensors on bicycles to monitor daily exercise or bike routes, and sensors in cars that can become available to either the phones or systems embedded in the cars to sense traffic, to detect road lanes, or to detect emergency situations. These external sensors are generally accessed through various wireless technologies (most prominently, Bluetooth is compatible with many external sensor devices). These external sensors will become more important in the future with the increase in the number and complexity of sensing applications.

2.2.4.3 Software APIs

The continuous addition of different new sensors in new smartphone models poses a challenge to smartphone operating systems, which must integrate the new sensors with the mobile applications. By default, smartphone operating systems provide basic sensor APIs to access the sensors from the mobile applications. However, these APIs provide access to raw data from the sensors, while leaving data analysis and noise filtering complexity to the application developers. Most developers do not want to handle this type of functionality in their applications. Instead, smartphone operating systems could provide higher-level sensor APIs to access filtered and context-based sensor data, which can be directly plugged into mobile applications such as activity recognition (e.g., walking, sitting, running, etc.).

Furthermore, smartphone energy consumption cannot be controlled if there are no limits over the level of sensor usage by third-party mobile applications. For example, if the mobile application continuously polls the GPS sensor just to get the city-level location accuracy, then the smartphone battery will not last very long. Instead, city-level accuracy can be achieved by using WiFi or cellular radios, which use less energy compared to GPS. Therefore, fine-grain controls over the sensors with support from a robust sensor API can help save the phone's battery power, and can enforce standards or best practices in accessing sensors in an efficient way by mobile application developers.

In recent years, smartphone operating systems have improved their sensor API support for third-party sensing applications to efficiently access the available smartphone sensors. However, the operating systems have not yet evolved to seamlessly support a sensing paradigm where sensor data collection, data aggregation, and context analysis can be done at the operating system level or even in a distributed fashion among co-located smartphones in a given area.

There are many research studies that proposed such sensing systems, but they have not yet been adopted by any operating system. Another challenge for developers programming sensing applications is that it is not easy to port their applications from one operating system to another, as different operating systems offer different sensor APIs to access the smartphone sensors. This is another reason why it is necessary to propose sensing abstractions and the standardization of sensor APIs.

Some of the prominent smartphone operating systems that provide sensor APIs are Google's Android, Apple's iOS, and Microsoft's Windows Mobile.

Google's Android APIs: Google's Android operating system, currently used by over 80% of the mobile devices worldwide, provides "Location and Sensors APIs" in the Java programming language. It uses sensors on the device to add rich context capabilities to mobile apps, ranging from GPS location to accelerometer, gyroscope, temperature, barometer, and more. Specifically, the Android framework location APIs are in the package *android.location*. In addition to these APIs, the *Google Location Services APIs* are available; they are part of Google Play Services (these services are available on all Android devices running Android version 2.3 or higher) and provide a more powerful, high-level framework that automates tasks such as location provider choice, Geofencing APIs, and power management. Location Services also provide new features, such as activity detection, that are not available in the original framework API. A new location service feature, called Fused Location Provider, intelligently manages the underlying location technology and gives the best location according to the needs of the sensing application.

The Android sensor framework lets the application developer access many types of sensors. The sensor framework is part of the *android.hardware* package. The most commonly used classes from the sensor's package are:

- SensorManager (used to create an instance of the sensor service).

- Sensor (used to create an instance of a specific sensor).

- SensorEvent (used to create a sensor event object, which provides information about a sensor event)

- SensorEventListener (used to create callback methods that receive notifications of sensor events when sensor values change or when sensor accuracy changes).

Some of the sensors available in Android devices are hardware-based and some are software-based. Hardware-based sensors get their data by directly measuring specific environmental properties, such as acceleration, geomagnetic field strength, or angular change. Software-based sensors are not physical devices, although they mimic hardware-based sensors. Software-based sensors derive their data from one or more of the hardware-based sensors and are sometimes called virtual sensors or synthetic sensors. The orientation sensor, linear acceleration sensor, and gravity sensor are examples of software-based sensors.

Apple's iOS APIs: The Apple's iOS APIs provides the *Core Motion Framework* in the Objective C programming language for accessing sensors. The Core Motion framework lets the sensing application receive motion data from the device hardware sensors and process that data. The framework supports accessing both raw and processed accelerometer data using block-based interfaces. For mobile devices with a built-in gyroscope, the app developer can retrieve the raw gyroscope data as well as the processed data reflecting the altitude and rotation rate of the device. The developer can use both the accelerometer and gyroscope data for sensing apps that use motion as input or as a way to enhance the overall user experience. The available sensor classes are CMAccelerometerData, CMAltitudeData, CMGyroData, CMMagnetometerData, CMPedometer, CMStepCounter, CMAltimeter, CMAttitude, CMMotionManager, and CMDeviceMotion.

Microsoft's Windows Mobile APIs: The Windows Mobile APIs provide the *Microsoft.Devices.Sensors* and *System.Device.Location* namespaces in the C++ and C# programming languages for accessing sensors. The Sensors namespace provides the Motion class, which enables applications to access information about the device's orientation and motion. Other available sensor classes are Accelerometer, Compass, and Gyroscope. The Location namespace provides access to the Windows Phone Location Service APIs, enabling the development of location-aware applications.

Basic Sensing Tasks: In a typical sensing application, the developer uses these sensor-related APIs to perform two basic tasks: 1) identifying sensors and sensor capabilities, and 2) monitoring sensor events. Identifying sensors and sensor capabilities at runtime is useful if the sensing application has features that rely on specific sensor types or capabilities. For example, the app developer may want to identify all of the sensors that are present on a device and disable any application feature that relies on sensors that are not present. Likewise, the app developer may want to identify all of the sensors of a given type so she can choose the sensor implementation that has the optimum performance for her sensing application. The app has to monitor sensor events to acquire raw sensor data. A sensor event occurs every time a sensor detects a change in the parameters it is measuring. A sensor event provides the app developer with four pieces of information: the name of the sensor that triggered the event, the timestamp of the event, the accuracy of the event, and the raw sensor data that triggered the event.

2.3 Where Are We?

The newest smartphones come with many embedded sensors, enabling a plethora of mobile sensing applications [33, 88, 44, 120] in gaming, smart environments, surveillance, emergency response, and social networks. The expanding sensing capabilities of mobile phones have gone beyond the original focus of sensor networks on environmental and infrastructure monitoring. Now, people are the carriers of sensing devices, and the sources and consumers of sensed data and events [149, 100, 119, 37, 108].

2.3.1 Current Uses

This section focuses on mobile sensing usage by presenting applications that take advantage of the available smartphone sensors and the phone software APIs to collect useful sensor data for both personal and public use.

Activity recognition through mobile sensing and wearable sensors has led to many *healthcare applications*, such as fitness monitoring [84], elder-care support, and cognitive assistance [54]. Wireless sensors worn by people for heart rate monitoring [9] and blood pressure monitoring [21] can communicate their information to the owners' smartphones. Phones are now able to monitor newborn jaundice through the "BiliCam" [61] application using the smartphone's camera; this app achieves high accuracy, comparable to the results from clinical tests, which require costly specialized equipment.

The sensing capabilities from smartphone sensors are becoming increasingly successful in the area of *road transportation*. Real-time traffic information can be collected from smartphones [23] such that the drivers can benefit from real-time traffic information. Parking data can be collected from cars equipped with ultra-sonic sensors [114]. The municipalities can quickly repair the roads by efficiently collecting pothole data using GPS and accelerometer sensors embedded in mobile devices [70].

Mobile sensing is also leveraged by the field of commerce for targeted *marketing/advertising*. The vendors/advertisers can use real-time location information collected from smartphones to target certain categories of people [11, 22] or can run context-aware surveys (as a function of location, time, etc.). For example, one question in such a survey could ask people attending a football game which other team games they would like to attend in the future.

Millions of people participate daily in online *social networks*, which provide a potential platform to utilize and share mobile sensing data. For example, the CenceMe project [119] uses the sensors in the phone to automatically sense the events in people's daily lives and selectively share this status on online social networks such as Twitter and Facebook, replacing manual actions that people now perform regularly.

Mobile sensing can be used in the *government sector* for measuring and reporting environmental pollution from a region or an entire city. Environment-protection agencies can use pollution sensors installed in phones to map with high accuracy the pollution zones around the country [26, 17]. The availability of ambient temperature sensors will soon enable the monitoring of weather from the smartphones for weather reporting organizations. Municipalities may collect data about phonic pollution, and then make an effort to reroute vehicular traffic at night from residential areas significantly affected by noise. Furthermore, governments in a few countries may soon insist on embedding radiation sensors [27] in all phones for detecting harmful radiation levels.

The cameras available on smartphones have improved greatly in recent years, and this allows news organizations to take advantage of these high-quality smartphone cameras to enable *citizen journalism* [4, 163, 25]. The citizens can report real-time data in the form of photos, videos, and text from public events or disaster areas. In this way, real-time information from anywhere across the globe can be shared with the public as soon as the events happen.

2.3.2   Known Hurdles

Mobile sensing performed on smartphones continuously for accurate activity recognition of the user or other sensing applications can lead to quick depletion of phone battery and can rise privacy issues. In addition, incentivizing the smartphone users to participate in the sensing is a major issue, as is the reliability of the sensing data from the participants.

Data reliability: The sensed data submitted by participants is not always reliable as they can submit false data to be compensated without executing the actual task. Therefore, it is important to validate the sensed data.

Incentives: It is not always economical for organizations (e.g., local/state government agencies or research communities) or individuals to provide monetary compensation for every type of sensing task that improves public services, helps with knowledge discovery, or improves our daily routine. Hence, it is necessary to find cost-effective solutions to incentivize the participants to perform sensing.

Battery consumption: In mobile application design and implementation, a significant issue faced by developers is battery consumption. Thus, controlling the actions of sensors and suspending them as needed is the key objective toward saving the phone's energy. For example, to achieve this goal, the developers and operating system APIs can apply the following standards: 1) use the operating system's energy profilers to detect the heavy usage of sensors when not needed and balance the quality of service and quality of user experience; 2) operating systems can provide APIs to cache the latest sensor data used, which might be sufficient for some sensing apps if the this data is in their specified threshold timeframe.

Privacy: Privacy is argued by most researchers to be the main hurdle facing mobile sensing [147]. When the sensing application uses third-party sensing services, the privacy is a major concern for the smartphone user because it is hard to trust the owners of servers where the sensor data is being uploaded. In addition, these servers can be attacked by hackers as well. However, the users feel safe as long as the sensed data remains on their phones. But without sending the sensor data to sensing services, the app developers may not be able to detect complex user activities. Furthermore, even if they could do it on the phone, this may result in high energy consumption or slow response time. Therefore, the app developers have to balance the application features provided to users with privacy.

The gravity of these sensing issues is discussed in detail in Chapters 7, 8, and 9. Existing solutions and newer advanced solutions that are just being proposed will be discussed.

2.4 Conclusion

In this chapter, we discussed the origin of mobile sensing from static WSNs to mobile people-centric sensing. We first reviewed the nature of static WSNs and the background of mobile sensing. We then presented the crucial components that enable mobile sensing. Subsequently, we presented the current uses of mobile sensing. Finally, we discussed the known hurdles facing mobile sensing in practice.

3

Crowdsourcing

3.1 Introduction

Crowdsourcing is the use of collective intelligence to solve problems in a cost-effective way. Crowdsourcing means that a company or institution has outsourced a task, which used to be performed by employees, to another set of people (i.e., crowd) [89]. Certain tasks could, of course, be done by computers in a more effective way. Crowdsourcing focuses on tasks that are trivial for humans, such as image recognition or language translation, which continue to challenge computer programs. In addition, crowdsourcing is used to perform tasks that require human intelligence and creativity.

People work on crowdsourcing tasks for payment, for the social good, or for other social incentives (e.g., competing against other people in a game). While the labor is not free, it costs significantly less than paying traditional employees. As a general principle, anyone is allowed to attempt to work on crowdsourcing tasks. However, certain tasks require expert knowledge (e.g., software development).

In addition to tasks done by individual users, crowdsourcing could employ groups or teams of users to perform complex tasks. For example, the book titled *The Wisdom of Crowds* [152] reveals a general phenomenon that the aggregation of information in groups results in decisions that are often better than those made by any single member of the group. The book identifies four key qualities that make a crowd smart: diversity of opinion, independence of thinking, decentralization, and opinion aggregation. This concept of crowd wisdom is also called "collective intelligence" [111].

In the rest of this chapter, we present the main categories of crowdsourcing applications, describe a number of crowdsourcing platforms that mediate the interaction between the task providers and task workers, and discuss open problems related to crowdsourcing such as recruitment, incentives, and quality control.

3.2   Applications

In the past decade, crowdsourcing has been used for many types of applications such as scientific applications, serious games/games with a purpose, research and development, commercial applications, and even public safety applications.

Some major beneficiaries of crowdsourcing are scientific applications. For example, Clickworkers [24] was a study that ran for one year to build an age map of different regions of Mars. Over 100,000 workers participated and they volunteered 14,000 work hours. Overall, the workers performed routine science analysis that would normally be done by scientists working for a very long time. It is important to notice that their analysis was of good quality: The age map created by workers agrees closely with what was already known from traditional crater counting.

Another project that employed crowdsourcing for science was Galaxy Zoo [8], which started with a data set made up of a million galaxies imaged by the Sloan Digital Sky Survey. The volunteers were asked to split the galaxies into ellipticals, mergers, and spirals, and to record the arm directions of spiral galaxies. Many different participants saw each galaxy in order to have multiple independent classifications of the same galaxy for high reliability classification. By the end of the first year of the study, more than 50 million classifications were received, contributed by more than 150,000 people. The scientists concluded that the classifications provided by Galaxy Zoo were as good as those from professional astronomers, and were subsequently used in many astronomy research papers.

While scientific applications rely purely on volunteers, other applications require the users to do some work in exchange for a service. For instance, reCAPTCHA [13] is a real-world service that protects websites from spam and abuse generated by bots (e.g., programs that automatically post content on websites). To be allowed to access a website, users need to solve a "riddle." In the original version of reCAPTCHA, the "riddle" took the form of deciphering the text in an image. The text was taken from scanned books, and thus crowdsourcing was used to digitize many books. Specifically, the service supplies subscribing websites with images of words that are hard to read for optical character recognition software, and the websites present these images for humans to decipher as part of their normal validation procedures. As the aggregate results for the same image converge, the results are sent to digitization projects. More recently, reCAPTCHA allows the users to perform image classification with mouse clicks. Hundreds of millions of CAPTCHAs are solved by people every day. This allows the building of annotated image databases and large machine learning datasets.

Serious games or games with a purpose represent another significant type of crowdsourcing applications. The workers in these games compete with each other to solve different problems. Similar to the newer version of reCAPTCHA, the ESP game [168] focuses on image annotation and classification. This online game randomly pairs players together and shows them the same image. The players do not know their partners and cannot communicate with each other. The goal of the game is to guess what label the partner would give to the image. Once both players have typed the exact same string, a new image appears. An image is considered classified after it passes through ESP multiple times and acquires several labels that people have agreed upon. During a four-month period, the game was played by over 13,000 people who generated over 1.2 million labels for about 300,000 images.

PhotoCity [164], another game with a purpose, is used to reconstruct real world locations as detailed 3D models from a large number of photos. The game exploits player mobility to collect carefully composed photographs of locations in an area of interest. Player photos include a wide variety of viewpoints necessary to build a complete reconstruction. To play, players go to locations in the real world that correspond to locations under construction in the virtual world and take photos. The game directs players to focus their attention on the gaps and fringes in a partial reconstruction in exchange for in-game rewards. The players compete for capturing virtual flags placed in real-world locations and taking ownership of familiar landmarks. Unlike other crowdsourced application, PhotoCity requires players to move in the physical world. Therefore, fewer players are expected to play the game, and this was proved by a user study that managed to have only 45 players. However, these players collected almost 100,000 images, which demonstrated the value of such an approach.

Crowdsourcing has succeeded as a commercial strategy for carrying work out as well, with companies accomplishing work ranging from crowdsourcing research and development to logo and t-shirt designs. InnoCentive [16] is one of the global leaders in crowdsourcing innovation problems. It provides challenge-driven and cost-effective innovation. For example, Forrester analyzed data from one customer and calculated that the organization achieved a return on investment of 74% [39]. People provide ideas and solutions to challenges in the fields of business, society, policy, science, and technology. Customers post challenges on InnoCentive along with the financial rewards to pay the solvers. Task solvers can choose and submit solutions to any of the challenges. Furthermore, they can work in groups. Once the customer ranks one solution as best, the solvers will be rewarded, the customer gains the intellectual property rights to the solution, and InnoCentive receives a percentage of the money involved in the transaction.

IdeaStorm [6] was launched by Dell as a platform to collect ideas for its new products and services. Users can post their ideas and suggestions on a certain product or service, and vote for promotion or demotion on the other ideas. Dell posts specific topics in "Storm Sessions" and asks customers to submit ideas. These Storm Sessions are open for a limited time in order to make the sessions relevant and time bound. Dell also added idea "Extensions," which enable comments posted on ideas to become part of the ideas themselves. Thus, ideas can evolve over time through collaboration. Once the idea is accepted, it is turned into reality by Dell. Since its inception, IdeaStorm has received over 16,000 ideas and implemented about 500 of them.

Crowdsourced software engineering [112] utilizes an open call format to recruit global online software engineers, to work on various types of software engineering tasks, such as requirements extraction, design, coding, and testing. This software development model can reduce the time-to-market by increasing parallelism, and lowering costs and defect rates. For example, TopCoder [29] states that crowdsourced development is capable of 30% to 80% cost savings compared to in-house development or outsourcing. In addition, the defect rates were reported to be 5 to 8 times lower compared with traditional software development practices. Yet another example, coming from Harvard Medical School, the best crowdsourced solution for DNA sequence-gapped alignment provided higher accuracy and was three times faster than the solution provided by the U.S. National Institute of Health [103].

The 99designs [1] company is the top marketplace for online web, logo, and graphics design contests. The customers build a design brief, pick a payment package, and share the design contest with a community of more than 1 million designers. The payment package is selected as a function of the number of designs to be received, the expertise of the designers, and whether there will be a dedicated manager for the project. Threadless [28] is a similar company that focuses on graphic designs for clothes, particularly t-shirts. Unlike 99designs, where the process is driven by customers, Threadless allows designers to post their t-shirt designs on the website where everybody else can buy them. Visitors vote on the designs, and Threadless selects the top 10 designs every week based on the voting results. These designs are then made available for customer purchase.

Public safety applications such as search and rescue operations or tracking stolen cars require a very large number of participants and, at the same time, fast execution. To test the potential of crowdsourcing to solve such problems, Defense Advanced Research Projects Agency (DARPA) launched its Network Challenge [132]. The challenge required teams to provide the coordinates of ten red weather balloons placed at different locations in the United States. According to the DARPA report, between 50 and 100 serious teams participated in the challenge and 350,000 people were involved. The MIT team won the challenge by completing it in 8 hours and 52 minutes. Advance preparation and good incentive mechanisms were the key to success, as the team recruited almost 4,400 participants in approximately 36 hours prior to the beginning of the challenge, through a recursive incentive mechanism.

3.3   Crowdsourcing Platforms

Crowdsourcing platforms manage the interaction between customers and workers. They act as trusted third parties that receive tasks from customers, distribute the tasks to workers, allow for team formation to perform the tasks, enforce contracts when the work is done, and maintain quality control throughout the process. The tasks received from customers have to be clearly specified such that the workers understand them easily. For complex tasks, the platform may help the customer by dividing them into smaller tasks that can be executed by individual workers. Many times, the tasks will be made available to all workers, who then self-select to work on them. However, if tasks have specific constraints such as reliability or deadlines, the platforms may need the quality of each worker and various completion time scenarios. Quality control can be achieved through methods such as voting, redundant workers, or worker reputation. Given the wide spectrum of crowdsourcing applications and their different requirements, it comes as no surprise that no single platform dominates the crowdsourcing market. This section reviews a few such platforms and identifies their similarities and differences.

Probably the best known crowdsourcing platform is as Amazon's Mechanical Turk (AMT) [3], which provides a market for human intelligence tasks (HIT) such as correcting spelling for search terms, choosing the appropriate category for a list of products, categorizing the tone of an article, or translating from English to French. AMT focuses on tasks that require human intelligence and cannot be solved easily by computers. Requesters post simple HIT tasks to be executed online by users around the world in exchange for micro-payments (on the order of a few cents to a few dollars). AMT ensures that workers are paid when the work is completed. At the same time, no worker can be paid until the requester confirms that the work is done. AMT receives a certain percentage of the money for every transaction. At the time of writing this book, AMT had over 1 million available tasks. One potential problem with AMT is that it is primarily used for simple, independent tasks.

CrowdForge [101] is a proposal for a general purpose platform for accomplishing complex and interdependent tasks using micro-task markets. Key challenges in such a platform include partitioning work into tasks that can be done in parallel, mapping tasks to workers, managing the dependencies between them, and maintaining quality control. CrowdForge provides a scaffolding for more complex human computation tasks that require coordination among many individuals, such as writing an article. CrowdForge abstracts away many of the programming details of creating and managing subtasks by treating partition/map/reduce steps as the basic building blocks for distributed process flows, enabling complex tasks to be broken up systematically and dynamically into sequential and parallelizable subtasks.

TopCoder [29] is a crowdsourcing platform for software designers, developers, and data scientists. It hosts weekly online competitions in algorithms, software design, and software development. The work in design and development results in software that is licensed for profit by TopCoder, which interacts directly with client companies to establish application requirements, timelines, budget, etc. Similar in nature to CrowdForge, TopCoder divides these complex application requirements into a set of components with a relatively small scope and expected behavior, including the interaction with other components. The set of requirements for all components is posted on the website as a single design competition. Any registered member can submit a design to any posted design competition. The winning design submissions become input for the development competition, which has similar structure. Once the competitors of this competition submit the code implementing the provided design, the output of the development competition is assembled together into a single application, which is later delivered to the customer. The workers not only receive money for their effort, but they also build up their resumes. Currently, TopCoder has over 7,000 competitions per year and a community of 1 million members who have been paid over $80 million.

Kaggle [18] is another online crowdsourcing platform for computing tasks, but it focuses on data mining and data science. Crowdsourcing allows large-scale and flexible invocation of human input for data gathering and analysis, which has broad applications in data mining. Kaggle has over 500,000 workers, who are paid and, at the same time, compete against each other. They can build networks to participate in team competitions, publish their scripts to improve their data science portfolio, and attract hiring managers with a strong community user ranking. CrowdFlower [5] is a similar system that combines machine learning and humans-in-the-loop in a single platform for data science teams doing sentiment analysis, search relevance, or business data classification.

Unlike platforms focused on specific categories of professional tasks, Witkeys [182] are a type of question-and-answer crowdsourcing platforms that are very popular in China. Users exchange and purchase services and information in order to save time and money. Generally, the posted questions are not easily answered by search engines and can be of either a personal or a professional nature. Taskcn is one of the largest Witkey platforms. On taskcn.com, users offer monetary rewards for solutions to problems, and other users compete to provide solutions in the hope of winning the rewards. The website plays the role of the third party by collecting the money from the requester and distributing it to the winner, who is decided by the requester. The website takes a small portion of the money as a service fee. Taskcn is similar to AMT to a certain extent, but it differs in two aspects. First, the requesters in AMT are generally companies, while in Taskcn they are other users. Second, workers always receive their payment once they complete the work in AMT. However, in Taskcn, users submit their work concurrently with other users competing on the same task, and they have no guarantee that their work will receive a monetary reward. Nevertheless, the site appears to be socially stable: There is a core of users who repeatedly propose and win. The large numbers of new users ensure many answers, while also providing new members for the stable core.

3.4   Open Problems

While crowdsourcing has been proven useful for solving many different types of tasks in a cost-effective way, it is still a niche activity in the global economic context. A first problem is how to break down a complex task to an ideal-sized work package suitable for the crowd. Currently, there are no standard efficient methods to achieve this goal, and therefore, crowdsourcing is mostly limited to relatively simple tasks.

A second problem is recruiting participants. Commercial platforms employ various types of open call formats, such as online competitions, on-demand matching, in which the workers are selected from the registrants, and online bidding, where the developers bid for tasks before starting their work. The lesson learned so far is that a combination of payment and social incentives (e.g., competitions, entertaining) works best. An interesting form of social incentive devised by successful crowdsourcing platforms is developing a professional reputation. The participants are able to gain experience in their professional field and measure up to others in their area. By earning points, they are able to develop a reputation that can enhance their resume and lead to hiring opportunities.

To recruit large crowds, one could employ the lessons learned from the DARPA Network Challenge. Large-scale mobilization toward a task requires diffusion of information about the task through social networks. Another lesson is the provision of incentives for individuals not only to work on the task but also to help recruit other individuals. Furthermore, the relation between demographics and user behavior needs to be studied in order to provide incentives appropriate for each type of community.

A related issue is how to make sure that the intended audience (e.g., level of education, prior training, or interest in the topic) for a crowdsourced task matches the intended outcome. The difficulty lies in controlling the crowd parameters to achieve outcomes with a certain level of consistency and statistical significance.

An essential problem in crowdsourcing is quality control, which includes cheating detection. Studies have shown that increased financial incentives increase the quantity, but not the quality, of the work [113]. In addition, higher payment could increase the cheating probability. Simple cheat-detection techniques are either based on control questions, evaluated automatically, or manually checked by the requester. If none of them is easily applicable, majority decision or using a control group to re-check tasks can be used. However, both solutions increase the cost per task.

Finally, other interesting problems in crowdsourcing include intellectual property protection, balancing the allocated budget for a task with the outcome quality, and project management automation.

3.5 Conclusion

This chapter presented an overview of crowdsourcing, with a focus on applications, platforms, and open issues. The examples described here demonstrate the success of crowdsourcing in a wide variety of domains from science to commerce and from software development to knowledge exchange. Crowdsourcing has an even bigger potential if its open problems such as recruitment, incentives, and quality control are solved. In this chapter, we also saw a few attempts to use crowdsourcing in the physical world (i.e., not online). These attempts represent a precursor to mobile crowdsensing, which is presented in in the next chapter.

4

What Is Mobile Crowdsensing?

4.1 Introduction

In Chapter 2, we have discussed two types of sensing: personal sensing and public sensing. Personal mobile sensing is favored mostly to monitor a single individual to provide customized alerts or assistance. Public mobile sensing, which is commonly termed as *mobile crowdsourcing*, requires the active participation of smartphone users to contribute sensor data, mostly by reporting data manually from their smartphones. The reports could be road accidents or new traffic patterns, taking photos for citizen journalism, etc. The aggregated useful public information can be shared back with the users on a grand scale.

The research advancements in crowdsourcing, and specifically mobile crowdsourcing, have set the course for mass adoption and automation of mobile people-centric sensing. The resulting new technology, as illustrated in Figure 4.1, is *"Mobile Crowdsensing."* This new type of sensing can be scalable and cost-effective for dense sensing coverage across large areas. In many situations, there is no need for organizations to own a fixed and expensive sensor network; they can use mobile people-centric sensing on demand and just pay for the actual usage (i.e., collected data).

This chapter looks first at mobile crowdsourcing and its advantages through the lens of illustrating applications. Then, it describes emerging crowdsensing applications in various domains and defines two types of crowdsensing, namely participatory manual sensing and opportunistic automatic sensing.

4.2 Advantages of Collective Sensing

The mobile crowdsourcing concept is similar to crowdsourcing, as commonly known from platforms such as Amazon's MTurk. This platform allows individuals and organizations (clients) to access a sheer number of people (participants) willing to execute simple sensing tasks for which they are paid. It allows smartphone users to collectively sense the data, which helps in the mon-

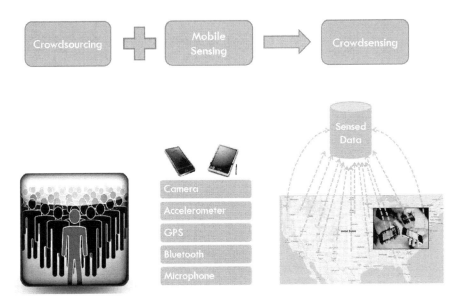

FIGURE 4.1
Mobile crowdsensing enables scalable and cost-effective sensing coverage of
large regions.

itoring of large-scale phenomena that cannot be easily measured by a single
individual.

For example, government organizations in metro cities may require traffic
congestion monitoring and air pollution level monitoring for better planning
of smart cities [26, 17]. This type of large-scale monitoring can be measured
precisely only when many individuals provide driving speed and air quality information from their daily commutes, which are then aggregated based
on location and time to determine traffic congestion and pollution levels in
the cities. Figure 4.2 shows the advantages of mobile crowdsourcing that are
prominently seen in the following domains: a) Environmental, b) Infrastructure, and c) Social.

4.2.1 Environmental

Mobile crowdsourcing helps improve the environment over large areas by measuring pollution levels in a city, or water levels in creeks, or monitoring wildlife
habitats. A smartphone user can participate in such applications to enable the
mapping of various large-scale environmental solutions. Common Sense [33] is
one such example, deployed for pollution monitoring. In the Common Sense
application, the participant carries the external air quality sensing device,
which links with the participant's smartphone using Bluetooth to record various air pollutants like $CO2$, NOx, etc. When large numbers of people col-

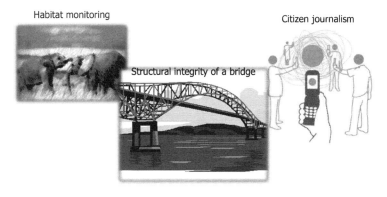

FIGURE 4.2
Advantages of mobile crowdsourcing seen in the environmental, infrastructure, and social domains.

lectively sense the air quality, it helps the environmental organizations take appropriate measures across large communities.

Similarly, the IBM Almaden Research Center had developed an application, "CreekWatch," to monitor the water levels and quality in creeks by collecting the data reported by individuals. The sensed data consist of pictures of various locations across the creeks and text messages about the trash in them. The water utility services can employ such data to track the pollution levels in the nearby water resources.

MobGeoSen [99] allows the participants to monitor the local environment for pollution and other characteristics. The application utilizes the microphone, the camera, the GPS, and external sensing devices (body wearable health sensors, sensors in vehicles, etc.). The application prompts the participants on the smartphone screen to add the text annotations and the location markers on the map during the participant's journey. The participants can share the photos with location and appropriate text tags, such that the collected data can be visualized easily on a spatial-temporal visualization tool. To accommodate the concurrent collection of sensing data from the external sensing devices, an advanced component was developed, which establishes multiple Bluetooth connections for communication. Furthermore, MobGeSen project works closely with science teachers and children at schools to demonstrate the use of the application by monitoring the pollution levels while traveling on their daily commutes to school.

NoiseTube [110] is a crowdsourcing application that helps in assessing the noise pollution by involving regular citizens. The participants in the application detect the noise levels they are exposed to in their everyday environment by using their smartphones. Every participant shares the noise measurements by adding personal annotations and location tags, which help to yield a collective noise map. NoiseTube requires participants to install the app on smartphones and requires a backend server to collect, analyze, and aggregate the data shared by the smartphones.

4.2.2   Infrastructure

The monitoring or measuring of public infrastructure can leverage the mobile crowdsourcing model, such that government organizations can collect status data for the infrastructure at low cost. Smart phone users participating in mobile crowdsourcing can report traffic congestion, road conditions, available parking spots, potholes on the roads, broken street lights, outages of utility services, and delays in public transportation. Examples of detecting traffic congestion levels in cities include MIT's CarTel [93] and Microsoft Research's Nericell [120]. The location and speed of cars is measured in the CarTel application by using special sensors installed in the car; these data are communicated to a central server using public WiFi hotspots. In Nericell, smartphones are used to determine the average speed, traffic delays, noise levels by honks, and potholes on roads.

Similarly, ParkNet [114] is an application that informs drivers about on-street parking availability using a vehicular sensing system running over a mobile ad hoc sensor network consisting of vehicles. This system collects and disseminates real-time information about vehicle surroundings in urban areas.

To improve the transportation infrastructure, the TrafficSense [121] crowdsourcing application helps in monitoring roads for potholes, road bumps, traffic jams, and emergency situations. In this application, participants use their smartphones to report the sensed data collected from various locations in their daily commutes. Furthermore, in developing countries, TrafficSense detects the honks from the vehicles using the audio samples sensed via the microphone.

PetrolWatch [66] is another crowdsourcing system in which fuel prices are collected using camera phones. The main goals of the system are to collect fuel prices and allow users to query for the prices. The camera lens is pointed toward the road by mounting it on the vehicle dashboard and the smartphone is triggered to capture the photograph of the roadside fuel price boards when the vehicle approaches the service stations. To retrieve the fuel prices, computer vision algorithms are used to scan these images. Each service station can have different style and color patterns on display boards. To reduce complexity, the computer vision algorithms are given the location information to know the service station and its style of the display board. While processing the image on the smartphone, the algorithms use the location coordinates, the brand, and the time. The prices determined after analyzing the image are uploaded to the central server and stored in a database that is linked to a GIS road network database populated with service station locations. The server updates fuel prices of the appropriate station if the current price has a newer timestamp. The system also retains the history of price changes to analyze pricing trends.

A pilot project in crowdsourcing, Mobile Millennium [86], allows the general public to unveil the traffic patterns in urban environments. These patterns are difficult to observe using sparse, dedicated monitoring sensors in the road infrastructure. The main goal of the project is to estimate the traffic on all major highways at specific targeted areas and also on the major interior city roads. The system architecture consists of GPS-enabled smartphones inside the vehicles, network provider, cellular data aggregation module, and traffic estimator. Each participant installs the application on the smartphone for collecting the traffic data and the backend server aggregates the data collected from all the participants. The aggregated data is sent to the estimation engine, which will display the current traffic estimates based on traffic flow models.

4.2.3 Social

There are interesting social applications that can be enabled by mobile crowdsourcing, where individuals share sensed information with each other. For example, individuals can share their workout data, such as how much time one exercises in a single day, and compare their exercise levels with those of the rest of the community. This can lead to competition and improve daily exercise routines. BikeNet [69] and DietSense [142] are crowdsourcing applications where participants share personal analytics with social networks or with a few private social groups. In BikeNet, smartphone users help in measuring the location and the bike route quality (e.g., route with less pollution, less bumpy ride) and aggregate the collected data to determine the most comfortable routes for the bikers. In DietSense, smartphone users take pictures of their lunch and share it within social groups to compare their eating habits. This is a good social application for the community of diabetics, who can watch what other diabetics eat and can even provide suggestions to others.

Party Thermometer [59] is a social crowdsourced application in which queries are sent to the participants who are at parties. For example, a query could be, "how hot is a particular party?" Similar to the citizen journalism applications, location is an important factor used to target the queries. But, unlike in the citizen journalism application, location alone is not enough for targeting because there is a significant difference between a person who is actually at a party and a person who is just outside, possibly having nothing to do with the party. Therefore, in addition to location, party music detection is also considered by employing the microphone to establish the user's context more accurately. To save energy on the phone, the sensing operations should happen only when necessary. Thus, the application first detects the location of the party down to a building, and only after that, it performs music detection using the microphone.

LiveCompare [63] is a crowdsourcing application that leverages the smartphone's camera to allow participants to hunt grocery bargains. The application utilizes a two-dimensional barcode decoding function to automatically identify grocery products, as well as localization techniques to automatically pinpoint store locations. The participants use their camera phones to take a photo of the price tag of their product of interest and the user's smartphone extracts the information about the product using the unique UPC barcode located on the tag. The price-tag barcodes in most grocery stores are identical to the barcodes on the actual products, which help in global product identification. The numerical UPC value and the photo are sent to LiveCompare's central server, once the barcode has been decoded on the smartphone. These data are stored in LiveCompare's database for use in future queries on price comparisons. The application provides high-quality data through two complementary social mechanisms to ensure data integrity: 1) it relies on humans, rather than machines, to interpret complex sale and pricing information; and 2) each query returns a subset of the data pool for a user to consider. The user can quickly flag it, if an image does not seem relevant. This lets users collectively identify malicious data, which can then be removed from the system.

4.3 How Can Crowdsourcing be Transformed into a Fun Activity for Participants?

Mobile crowdsourcing is beneficial and efficient for people when they participate in large numbers. But it is not easy to motivate the smartphone users to share the sensed data for social good. Monetary incentives are one way to improve participation, but it may sometimes lead to high costs when sensing is required from large areas. Therefore, if these crowdsourcing applications are built as fun activities among participants, then the users will be more excited to use them, and thus collect the required amount of sensing data. Crowdsourcing applications can be transformed into fun games in which participants are challenged to play the game and to score the highest points.

In one crowdsourced game [128], the mobile users have to record as many audible signals as possible from different traffic lights. This collected data can be processed and integrated with Google maps to improve crossroad accessibility for blind pedestrians by providing information about the accessibility of their routes. The players receive points for each data point they collect as a function of the contribution is trustworthiness and originality. Such a crowdsourced application motivated other researchers to apply the same gaming concept to collect a wide variety of sensing data on a large scale.

BioGame [115] shows that in cases where medical diagnosis is a binary decision (e.g., positive vs. negative, or infected vs. uninfected), it is possible to make accurate decisions by crowdsourcing the raw data (e.g., microscopic images of specimens/cells) using entertaining digital games (i.e., BioGames) that are played on PCs, tablets, or mobile phones. This work mainly focuses on the problem of handling large quantities of data through crowdsourcing.

BudBurst [81] is a smartphones application for an environmental participatory sensing project that focuses on observing plants and collecting plant life-stage data. The main goal is "floracaching," for which players gain points and levels within the game by finding and making qualitative observations on plants. This game is also an example of motivating participatory sensing.

Another participatory sensing game, Who [80], is used to extract relationships and tag data about employees. It was found useful for rapid collection of large volumes of high-quality data from "the masses."

4.4 Evolution of Mobile Crowdsensing

A formal definition of mobile crowdsensing [79] is that it is a new sensing paradigm that empowers ordinary citizens to contribute data sensed or generated from their mobile devices. The data is then aggregated and fused in the cloud for crowd intelligence extraction and people-centric service delivery. From an artificial intelligence (AI) perspective, mobile crowdsensing is founded on a distributed problem-solving model where a crowd is engaged in solving a complex problem through an open call [41]. As detailed in Chapter 3, the concept of crowd-powered problem-solving has been explored in several research areas in the past. Different from the the concepts that focus on the advantages of group decision making, mobile crowdsensing is mainly about the crowd-powered data collection and analysis process.

Recently, several mobile crowdsourcing projects have tried to leverage traditional crowdsourcing platforms for crowdsensing. Twitter [31] has been used as a publish/subscribe medium to build a crowdsourced weather radar and a participatory noise-mapping application [62]; mCrowd [179] is an iPhone-based platform that was used to build an image search system for mobile phones, which relies on Amazon MTurk [3] for real-time human validation [180]. These solutions have the advantage of leveraging the popularity of existing crowdsourcing platforms (tens of thousands of available workers), but do not allow for truly mobile sensing tasks to be performed by workers (i.e., tasks that can only be performed using sensors on mobile phones). Moreover, as explained in Chapter 2, an autonomous mobile crowdsensing platform introduces additional concerns that must be addressed, such as the privacy of the participants and the reliability of the sensed data.

4.4.1 Emerging Application Domains

In the following, we present several domains that can benefit from mobile crowdsensing, as well as a number of applications (some of them already prototyped) for each domain:

- *Road Transportation*: Departments of transportation can collect fine-grain and large-scale data about traffic patterns in the country/state using loca-

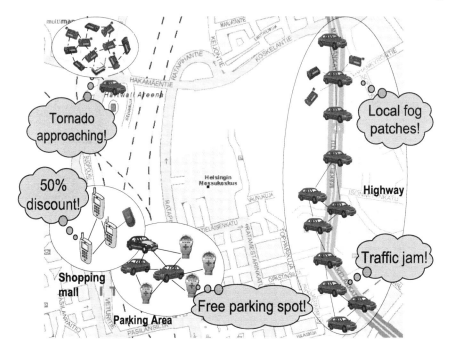

FIGURE 4.3
Mobile crowdsensing: People are both consumers and providers of sensed data.

tion and speed data provided by smartphones. The same information could be used to provide individualized traffic re-routing guidance for congestion avoidance [129] or to direct drivers toward free parking spots [114]. Data about the quality of the roads could also be collected to help municipalities quickly repair the roads [70]. Similarly, photos (i.e., camera sensor data) taken by people during/after snowstorms can be analyzed automatically to prioritize snow cleaning and removal.

- *Healthcare and Wellbeing*: Wireless sensors worn by people for heart rate monitoring [9] and blood pressure monitoring [21] can communicate their information to the owners' smartphones. Typically, this is done for both real-time and long-term health monitoring of individuals. Crowdsensing can leverage these existing data into large-scale healthcare studies that seamlessly collect data from various groups of people, which can be selected based on location, age, etc. A specific example involves collecting data from people who regularly eat fast food. The phones can perform activity recognition and determine the level of physical exercise done by people, which was proven to directly influence people's health. For example, as a result of such a study in a city, the municipality may decide to create more bike lanes to encourage people to do more physical activities.

Similarly, the phones can determine the level of social interaction of certain groups of people (e.g., using Bluetooth scanning, GPS, or audio sensors). For example, a university may discover that students (or students from certain departments) are not interacting with each other enough; consequently, it may decide to organize more social events on campus. The same mechanism coupled with information from "human sensors" can be used to monitor the spreading of epidemic diseases.

- *Marketing/Advertising*: Real-time location or mobility traces/patterns can be used by vendors/advertisers to target certain categories of people [11, 22]. Similarly, they can run context-aware surveys (as a function of location, time, etc.). For example, one question in such a survey could ask people attending a concert what artists they would like to see in the future.

4.4.2 Crowdsensing in Smart Cities

The power of mobile crowdsensing can be applied in smart cities. The main visionary goal is to automate the organization of spontaneous and impromptu collaborations of large groups of people participating in collective actions (such as ParticipAct [50]) in urban crowdsensing. In a crowdsensing environment, people or their mobile devices act as both sensors that collect urban data and actuators that take actions in the city, possibly upon request. Managing the crowdsensing process is a challenging task spanning several socio-technical issues: from the characterization of the regions under control to the quantification of the sensing density needed to obtain a certain accuracy; from the evaluation of a good balance between sensing accuracy and resource usage (number of people involved, network bandwidth, battery usage, etc.) to the selection of good incentives for people to participate (monetary, social, etc.). SmartSantander [161] is an example of a large-scale participatory sensing infrastructure using smartphones deployed within a Smart City.

From a social perspective, there is the need to identify people willing to participate in urban sensing tasks and to find good incentives for participation, not only monetary rewards but also social ones (e.g., cleaner and safer cities). Once identified, the participants have to be kept in the crowdsensing loop, which involves active participation in sensing campaigns. From a more technical perspective, one of the main challenges is finding a good balance between system scalability and sensing accuracy for city-wide deployment environments. In such a new socio-technical system, the types of resources are very different, spanning from computing (network bandwidth, memory, CPU, etc.) to human resources (number of people involved, human attention, personal skills to contribute, etc.). Thus, it is impossible to fully control them.

4.5   Classification of Sensing Types

The mobile crowdsensing model treats each person carrying a mobile device as a potential sensor node and forms a large sensor network leveraging the entire population. Crowdsensing can be broadly classified into two major categories:

- Participatory manual sensing, where the participant manual intervention is needed for certain input.

- Opportunistic automatic sensing, where the sensing is performed on the participant's smartphones in the background.

4.5.1   Participatory Manual Sensing

In participatory sensing, each sensing task involves the participant directly to complete certain actions, such as capturing the photos of locations or events. Therefore, in this method, the participants have major control of when, where, and what to sense by deciding which sensing tasks to accept and to complete, based on their availability. In participatory sensing, it is possible that sensing is done automatically, but the participants may control the time when the sensing can be done, the sharing policies, etc. In the following, we discuss examples of participatory sensing applications and extract crowdsensing insights from these applications.

Photo and video journalism In *citizen journalism* [163, 25] applications, citizens can report real-time data in the form of photos, video, and text from public events or disaster areas. In this way, real-time information from anywhere across the globe can be shared with the public as soon as the event happens.

MoVi [38] is a crowdsensing system that creates video highlights based on videos recorded by multiple people at real-life social gatherings and parties. While this application assumes that participants record videos continuously, we believe a more realistic scenario assumes that videos are recorded only during certain events. Interesting events among various social groups are identified by a detection module. Once an interesting event is detected in a group, the smartphones's data of each participant is analyzed at a server to pick the best video for the event. The extracted best video clips are then sorted in time and stitched into a single video comprising the highlights of the interesting event. The application running at the server automatically does all the processing in generating such a video.

Road transportation sensing In *traffic-jam alert* [85, 175] applications, the Department of Transportation uses a mobile crowdsensing system to collect alerts generated by people driving on congested roads and then distributes these alerts to other drivers. In this way, drivers on the other roads can benefit from real-time traffic information.

Data sharing in social networks PEIR [124] is a participatory crowdsensing application that uses location data collected from smartphones to calculate the personalized approximations of environmental effect and exposure. The application collects and uploads data automatically to the server where the data is processed and analyzed. The application's web interface provides information to the participants, such as Carbon Impact, Sensitive Site Impact, Smog Exposure, and Fast Food Exposure, which help to determine how individual choices influence both the environmental impact and exposure. The manual part of this application happens when the participants share and compare their data or reports with other participants or their friends in the social networks.

A similar combination of automatic sensing and manual sharing is encountered in another participatory crowdsensing application, CenceMe [119]. People can share their daily activities on social networks through smartphone sensing. The CenceMe application detects participants' activities, such as walking, running, or sitting; their current mood such as sad, happy, or angry; and also the daily visited locations such as daily gym, regular coffee shop, and workplace. The participants decide on which social networks to post their data. This application allows the participants to be active among their friends and helps them in finding new social connections. The CenceMe software is comprised of: a) software-based sensors and data analysis engine to perform activity recognition; b) sensing data repository to store the collected data; and c) a service layer used to share the sensed data between friends based on participants' privacy settings.

4.5.2   Opportunistic Automatic Sensing

In opportunistic sensing, the actual sensing is performed automatically and the participant is not involved in the sensing activity. The smartphones makes the appropriate decisions and initiates the sensing data collection and sharing. This sensing model transfers the burden from the participants to a central sensing system (or to an ad hoc network of mobile devices) to determine when, where, and on which participant's smartphones to perform automatic sensing in order to satisfy the sensing application needs. The system considers the social aspects of the participants when making decisions regarding sensing task scheduling. These aspects include current location, possible next location, current activity, regular patterns on a daily basis, and health condition; then this sensing aspect is called people-centric sensing.

Behavioral and social sensing Darwin [118] is an opportunistic crowdsensing platform that combines collaborative sensing and classification techniques to detect participants' behavior and their context on mobile phones. The system leverages efficient machine learning techniques specially built to run directly on smartphones with limited storage and processing resources. Many social networking applications can easily integrate Darwin. The application focuses on voice recognition (i.e., the voice is captured by the microphones) and executes three important steps: 1) each smartphones constructs a model of the event to be sensed, and over time, the model evolves with new data; 2) multiple smartphones that are co-located transfer their models to each other to enable classification tasks for different people, and 3) the classification accuracy is achieved by collaborative inference of various smartphones sensing context viewpoints. The raw sensor data is never communicated to other smartphones or stored on the smartphone, in order to improve the application's performance and privacy.

EmotionSense [135] is another mobile opportunistic sensing platform for social psychological studies. The application senses individual emotions, activities, and interactions between the participants of social groups. The application can be used to understand human behavior and emotions by correlating participants' interactions and activities. The main novelties in EmotionSense are the emotion detection and speaker recognition modules, which are based on Gaussian mixture methods. The experimental results in the real world, done in collaboration with social psychologists, show that the distribution of the emotions detected through EmotionSense generally reflects self-reports by the participants.

Road transportation sensing Drivers looking for parking cause on average, 30% of traffic congestion in urban areas as per automotive traffic studies. However, deploying a sensing infrastructure to monitor real-time parking availability throughout a city is a significant investment burden on city transportation organizations. Instead, ParkSense [126] proposes a system that detects if a driver has vacated a parking spot. In conjunction with crowdsensing, this system can be leveraged to address many parking challenges. ParkSense leverages WiFi signatures in urban areas for sensing "unparking" events. In addition, it uses a novel approach based on the rate of change of WiFi beacons to sense if the person has started driving.

Mobile crowdsensing can be used to provide a queue system for taxi services, similar to passengers waiting queues and taxis waiting queues that are commonly seen in many urban cities. In densely populated cities, either passengers queuing for taxis or taxis queuing for passengers frequently occurs due to the citywide imbalance of taxi supply and demand. Timely and accurate detection of such waiting queues and the queue properties would undoubtedly benefit both public commuters and taxi drivers. A queue analytics system has been proposed in [109], one that employs crowdsensing to collect mobile data from taxis and smartphones in order to detect both passenger queues and taxi queues. In particular, the system first determines the existence of taxi queues by analyzing the taxi data, and then makes a soft inference on passenger queues. On the taxi side, taxis periodically update their status, GPS location, and instantaneous speed. Meanwhile, the passenger side adopts a crowdsensing strategy to detect the personal-scale queuing activities. The extensive empirical experiments demonstrated that the system can accurately and effectively detect the taxi queues and then validate the passenger queues.

Indoor localization In mobile applications, location-based services are becoming increasingly popular to achieve services such as targeted advertisements, geosocial networking, and emergency notifications. Although GPS provides accurate outdoor localization, it is still challenging to accurately provide indoor localization even by using additional infrastructure support (e.g., ranging devices) or extensive training before system deployment (e.g., WiFi signal fingerprinting). Social-Loc [97] is designed to improve the accuracy of indoor localization systems with crowdsensing. Social-Loc takes as its input the potential locations of individual users, which are estimated by any underlying indoor localization system, and exploits both social encounters and non-encounter events to cooperatively calibrate the estimation errors. Social-Loc is implemented on the Android platform and demonstrated its performance over two underlying indoor localization systems: Dead-reckoning and WiFi fingerprint.

Furthermore, in most situations the lack of floor plans makes it difficult to provide indoor localization services. Consequently, the service providers have to go through exhaustive and laborious processes with building operators to manually gather such floor-plan data. To address such challenges, Jigsaw [71], a floor-plan reconstruction system, is designed such that it leverages crowdsensed data from mobile users. It extracts the position, size, and orientation information of individual landmark objects from images taken by participants. It also obtains the spatial relation between adjacent landmark objects from inertial sensor data and then computes the coordinates and orientations of these objects on an initial floor plan. By combining user mobility traces and locations where images are taken, it produces complete floor plans with hallway connectivity, room sizes, and shapes.

4.6 Conclusion

This chapter discussed the origins of mobile crowdsensing, which borrows techniques from people-centric mobile sensing and crowdsourcing. We first discussed the advantages of collective sensing in various domains such as environmental, infrastructure, and social. Then, we investigated non-monetary incentives for mobile sensing, with a focus on gaming. The chapter continued with an extensive presentation of mobile crowdsensing and its application domains. Finally, we presented a classification of sensing types in mobile crowdsensing and illustrated each type with representative applications.

5

Systems and Platforms

5.1 Introduction

This chapter describes several mobile crowdsensing systems that are based on a centralized design, such as McSense [156], Medusa [134], and Vita [92]. The chapter also describes the prototype implementation and the sensing tasks developed for each mobile crowdsensing system.

5.2 The McSense System

McSense [156] is a mobile crowdsensing platform that allows clients to collect many types of sensing data from smartphones carried by mobile users. The interacting entities in the mobile crowdsensing architecture are:

- *McSense:* A centralized mobile crowdsensing system that receives sensing requests from clients and delivers them to providers.

- *Client:* The organization or group that is interested in collecting sensing data from smartphones using the mobile crowdsensing system.

- *Provider:* A mobile user who participates in mobile crowdsensing to provide the sensing data requested by the client.

5.2.1 Architecture of McSense

The architecture of McSense, illustrated in Figure 5.1, has two main components: (1) the server platform that accepts tasks from clients and schedules the individual tasks for execution at mobile providers; and (2) the mobile platform (at the providers) that accepts individual tasks from the server, performs sensing, and submits the sensed data to the server. The communication among all these components takes place over the Internet. Next we discuss the overall process in more detail.

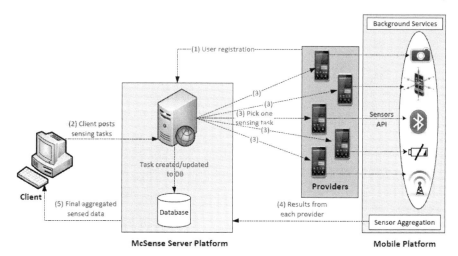

FIGURE 5.1
McSense architecture.

User registration: The McSense application on the smartphone shows a registration screen for first-time users, prompting them to enter an email address and a password. During the registration process, the user phone's MEID (Mobile Equipment IDentifier) is captured and saved in the server's database along with the user's email address and password. The authors chose to store the phone's MEID in order to enforce one user registration per device. In addition, the server also avoids duplicate registrations when users try registering with the same email address again.

Posting new sensing tasks: New sensing tasks can be posted by clients using a web interface running on the McSense server. The sensing task details are entered on this web page by the client and submitted to the server's database. Once a new task is posted, the background notification service running on the provider's phone identifies the new available tasks and notifies the provider with a vibrate action on the phone. Providers can check the notification and can open the McSense application to view the new available tasks. When the application is loaded, the providers can see four tabs (Available, Accepted, Completed and Earnings). The providers can view the list of tasks in the respective tabs and can click on each task from the list to view the entire task details (type, status, description, accepted time, elapsed time, completion time, expiration time, payment amount).

Life cycle of a task: The life cycle starts from the Available tasks tab. When a provider selects an available task and clicks on the Accept button, the task is moved to the Accepted tab. Once a task is accepted, then that task is not available to others anymore. When the accepted task is completed according to its requirements, the task is moved to the Completed tasks tab. Finally, the providers view their aggregated total dollars earned for successfully completed tasks under the Earnings tab. If the accepted task expires before being completed successfully according to its requirements, it is moved to the Completed tasks tab and marked as unsuccessfully completed. The providers do not earn money for the tasks that are completed unsuccessfully.

Background services on phone: When the network is not available, a completed task is marked as pending upload. A background service on the phone periodically checks for the network connection. When the connection becomes available, the pending data is uploaded and finally, these tasks are marked as successfully completed. If the provider phone is restarted manually or due to a mobile OS crash, then all the in-progress sensing tasks are automatically resumed by the Android's BroadcastReceiver service registered for the McSense application. Furthermore, the Accepted and the Completed tabs' task lists are cached locally and are synchronized with the server. If the server is not reachable, the users can still see the tasks that were last cached locally.

5.2.2 Tasks Developed for McSense

The sensing tasks that were developed fall into two categories:

1. Manual tasks, e.g., photo tasks.
2. Automated tasks, e.g., sensing tasks using accelerometer and GPS sensors; sensing tasks using Bluetooth communication.

Manual Photo Sensing Task: Registered users are asked to take photos from events on campus. Once the user captures a photo, she needs to click on the "Complete Task" button to upload the photo and to complete the task. Once the photo is successfully uploaded to the server, the task is considered successfully completed. These uploaded photos can be used by the university news department for their current news articles. On clicking the "Complete Task" button, if the network is not available, the photo task is marked as completed and waiting for upload. This task is shown with a pending icon under the completed tasks tab. A background service would upload the pending photos when the network becomes available. If a photo is uploaded to the server after the task expiration time, then the photo is not useful to the client. Therefore, the task will be marked as "Unsuccessfully completed," and the user does not earn money for this task.

Automated Sensing Task using Accelerometer and GPS Sensors: The accelerometer sensor readings and GPS location readings are collected at 1-minute intervals. The sensed data is collected along with the userID and a timestamp, and it is stored into a file in the phone's internal storage, which can be accessed only by the McSense application. This data will be uploaded to the application server on completion of the task (which consists of many data points). Using the collected sensed data of accelerometer readings and GPS readings, one can identify users' activities like walking, running, or driving. By observing such daily activities, one can find out how much exercise each student is getting daily and derive interesting statistics, such as which department has the most active and healthy students in a university.

Automated Sensing Task using Bluetooth Radio: In this automated sensing task, the user's Bluetooth radio is used to perform periodic (*every 5 minutes*) Bluetooth scans until the task expires; upon completion, the task reports the discovered Bluetooth devices with their location back to the Mc-Sense server. The sensed data from Bluetooth scans can provide interesting social information, such as how often McSense users are near each other. Also, it can identify groups who are frequently together, to determine the level of social interaction among certain people.

Automated Resource Usage Sensing Task: In this automated sensing task, the usage of a user's smartphone resources is sensed and reported back to the McSense server. Specifically, the report contains the mobile applications usage, the network usage, the periodic WiFi scans, and the battery level of the smartphone. While logging the network usage details, this automated task also logs overall device network traffic (transmitted/received) and per-application network traffic.

5.2.3 Field Study

The McSense platform was used to perform a field study. To participate in the study, students have been asked to download the McSense application from the Android market and install it on their phones. The demographic information of the student participants is shown in Table 5.1. On the application server, various sensing tasks are posted periodically. Some tasks had high monetary value associated with their successful completion. A few other tasks were offered with very low or no monetary incentives, in order to observe the provider's participation (i.e., volunteering) in collecting free sensing data. As tasks are submitted to the application server, they also appear on the participants' phones where the application has been installed. Each task contains a task description, its duration, and a certain amount of money. The students use their phones to sign up to perform the task. Upon successful completion of the task, the students will accumulate credits (payable in cash at the end of the study). The authors conducted the study for approximately 2 months.

Next, provide details about the automated accelerometer and GPS sensing tasks posted on the McSense server:

1. The authors posted automated tasks only between 6am and 12pm. Users can accept these tasks when they are available. When a user accepts the automated sensing task, then the task starts running in the background automatically. Furthermore, to be able to accept

TABLE 5.1

Participants demographic information

Total participants	58
Males	90%
Females	10%
Age 16–20	52%
Age 21–25	41%
Age 26–35	7%

automated sensing tasks, users must have WiFi and GPS radios switched on.

2. Automated sensing tasks expire each day at 10pm. The server compares the total sensing time of the task to a threshold of 6 hours. If the sensing time is below 6 hours, then the task is marked as "Unsuccessfully Completed," otherwise it is marked as "Successfully Completed."

3. Automated sensing tasks always run as a background service. On starting or resuming this service, the service always retrieves the current time from the server. Thus, even when the user sets an incorrect time on the mobile, the task will always know the correct current time and will stop sensing after 10pm.

4. Long-term automated sensing tasks are posted for multiple days. Users are paid only for the number of days they successfully complete the task. The same threshold logic is applied to each day for these multi-day tasks.

For manual tasks such as photo tasks, users have to complete the task manually from the Accepted Tasks Tab by taking the photo at the requested location. Users were asked to take general photos from events on a university campus. Once the photos are successfully uploaded to the application server and a basic validation is performed (photos are manually validated for ground truth), the task is considered successfully completed.

5.3 The Medusa System

Medusa [134] is a programming framework for crowdsensing that provides support for humans-in-the-loop to trigger sensing actions or to review results. It recognizes the need for participant incentives, and addresses their privacy and security concerns. Medusa provides high-level abstractions for specifying the steps required to complete a crowdsensing task, and employs a distributed runtime system that coordinates the execution of these tasks between smartphones and a cluster on the cloud. The Medusa architecture and design is evaluated using a prototype that uses ten crowdsensing tasks.

5.3.1   Architecture of Medusa

Medusa addresses the challenge of programming crowdsensing tasks elegantly by using a high-level programming language called "MedScript." This language can reduce the burden of initiating and managing crowdsensing tasks, specifically for non-technical requestors who will constitute the majority of crowdsensing users. MedScript provides abstractions for intermediate steps in a crowdsensing task (these are called "stages"), and for controlling the flow between stages (these are called "connectors"). Unlike most programming languages, it provides the programmer with language-level constructs for incorporating workers into the sensing workflow. The Medusa runtime is architected as a partitioned system with a cloud component and a smartphone component. For robustness, it minimizes the task execution state maintained on the smartphone, and for privacy, it ensures that any data that leaves the phone must be approved by the (human) worker. Finally, it uses Amazon Mechanical Turk(AMT) to recruit workers and to manage monetary incentives.

The design of the Medusa runtime is guided by three architectural decisions that simplify its overall system design:

- Partitioned Services: The Medusa design is focused on implementing a partitioned system that uses a collection of services both on the cloud and on worker smartphones. The intuition behind such an architectural decision is that some of the requirements are more easily accomplished on an always Internet-connected cloud server or cluster (such as task initiation, volunteer recruitment, result storage, and monetary transactions). Others, such as sensing and in-network processing, are better suited for execution on the smartphone.

- Dumb Smartphones: The sensing system should minimize the amount of task execution state that is maintained on smartphones. This design principle prevents large segments of a task from being completely executed on the smartphone without any task progress updates sent to the cloud. This principle enables robust task execution even in the case of intermittent connectivity, failures, or long-term phone outages that may occur when workers turn off their smartphones.

- Opt-in Data Transfers: The sensing system should automatically require a user's permission before sending any data from a smartphone to the cloud. Data privacy is a significant concern in crowdsensing applications, and this principle ensures that the users have the option to opt-out of data contributions. Furthermore, before the workers opt-in to share sensing data, they may view a requestor's privacy policy.

These architectural principles enable the mobile crowdsensing system to achieve all of the requirements of the requestors (sensing clients) without significantly complicating the system design.

5.3.2 Execution of Sensing Tasks in Medusa

According to the partitioned services principle, Medusa is structured as a collection of services running on the cloud and on the phone. These services coordinate to perform crowdsensing tasks. In this section, we give a high-level overview of Medusa by using a running example about a user's video documentation task (i.e., the user is collecting relevant videos for her research).

In Medusa, Alice writes her video documentation task in MedScript. She submits the program to the MedScript interpreter, which is implemented as a cloud service. The sequence of stages specified in the video documentation crowdsensing task is as follows:

$Recruit-> TakeVideo-> ExtractSummary-> UploadSummary->$
$Curate-> UploadVideo$

The interpreter parses the program and creates an intermediate representation, which is passed on to a Task Tracker. The Task Tracker coordinates the task execution with other components as well as with workers' smartphones. In the Recruit stage, Task Tracker contacts Worker Manager (a back-end service), which initiates the recruitment of workers (it uses Amazon Mechanical Turk for recruitment). When workers agree to perform the task, these notifications eventually reach Task Tracker through Worker Manager. Generally, different workers may agree to perform the task at different times. Monetary incentives are specified at the time of recruitment.

Once a worker has been recruited, the next stage, called TakeVideo, is initiated on that worker's smartphone by sending a message to the Stage Tracker instance that runs on every phone. The TakeVideo stage is a downloadable piece of code from an extensible Stage Library, a library of pre-designed stages that requestors can use to design crowdsensing tasks. Each such stage executes on the phone in a sandboxed environment called MedBox. The TakeVideo stage requires human intervention where the worker needs to open the camera application and take a video. To remind the worker of this pending action, the stage implementation uses the underlying system's notification mechanism to alert the worker. Once the video has been recorded, Stage Tracker notifies the Task Tracker of the completion of that stage, and awaits instructions regarding the next stage, which adheres to the dumb smartphones principle.

The Task Tracker then notifies the Stage Tracker that it should run the ExtractSummary stage. This stage extracts a small summary video comprised of a few sample frames from the original video, then uploads it to the Task Tracker. Before the upload, the user is required to preview and approve the contribution as per the opt-in data transfers principle.

The execution of the next stages follows the same pattern. The Task Tracker initiates the Curate stage, in which a completely different set of volunteers rates the videos. Finally, the selected videos are uploaded to the cloud. Finally, Alice is notified once all the stages are completed.

5.3.3   Scalability and Robustness

To analyze the scalability and overhead in Medusa, the system measured the time taken to perform several individual steps in task execution, both on the cloud and the smartphones. The main server-side delay in Medusa is due to the time that it takes to deliver an SMS message to the phone: between 20 and 50 seconds. This is because the implementation uses an email-to-SMS gateway; a more optimized implementation might use direct SMS/MMS delivery, reducing latency. However, latency of task execution is not a significant concern when humans are in the loop. For crowdsensing tasks it would be better to have deadlines on the order of hours from initiation. In addition, the second main component of delay in Medusa is the time that it takes for a human to sign up for the task.

Overall, actual processing overheads on the server side are only 34.47ms on average, which permits a throughput of 1740 task instances per minute on a single server. Besides, the server component is highly parallelizable, since each task instance is relatively independent and can be assigned a separate Task Tracker. Therefore, if task throughput ever becomes an issue, it would be possible to leverage the cloud's elastic computing resources to scale the system.

The major component that adds delay on the smartphone is the time to instantiate the stage implementation binary, which takes 402.7ms. This overhead is imposed by Android as the stage implementations are Android package files and the Android runtime must unpack this and load the class files into the virtual machine. Finally, the total delay imposed by Medusa platform is only 459.8ms per stage, of which the dominant component is the stage binary initialization. On the timescale of expected task completion (generally in the order of 10s of minutes) this represents negligible overhead.

To enable robust task execution, Medusa maintains the state of data contributions in its Data Repository as well as the state of stage execution for each worker in its Stage State Table. In case of a task failure, Medusa waits for a pre-defined timeout (10 minutes in the prototype implementation) before restarting the pending stage of the sensing task. This simple retry mechanism works well across any transient failure mainly because stage execution state is maintained on the cloud. Additionally, Medusa allows smartphone owners to limit the computation used by a single stage (on a per-stage basis) and can also limit the amount of data transferred per task instance. This allows users to implement policies that allocate more resources for tasks that get a higher monetary reward.

5.4 The Vita System

Vita [92] is a mobile cyber-physical system for crowdsensing applications, which enables mobile users to perform mobile crowdsensing tasks in an efficient manner through mobile devices. A cyber-physical system (CPS) is a system of collaborating computational elements controlling physical entities. A mobile cyber-physical system is a subcategory of CPS. Vita provides a flexible and universal architecture across mobile devices and cloud computing platforms by integrating the service-oriented architecture with a resource optimization mechanism for crowdsensing, with extensive support to application developers and end users. The customized platform of Vita enables intelligent deployments of tasks between humans in the physical world. It also allows dynamic collaborations of services between mobile devices and a cloud computing platform during run-time of mobile devices with service failure handling support.

5.4.1 Architecture of Vita

The two main components of Vita's architecture are the mobile platform and the cloud platform. The mobile platform provides the required environment and ubiquitous services to allow users to participate in crowdsensing tasks through their smartphones. The cloud platform provides a central coordinating platform to store and integrate the wide variety of data and tasks from crowdsensing and social networking service providers, as well as the development environment to support the development of mobile crowdsensing applications.

Vita uses the **mobile service-oriented architecture** (SOA) framework. This is an extensible and configurable framework that is based on the specifications and methodologies of RESTful Web Services. The mobile SOA integrates popular social networking services such as Facebook and Google+. It also supports the development of multiple mobile web service-based applications and services in an efficient and flexible way, with standard service interaction specifications that enable dynamic service composition during mobile devices' run-time. Furthermore, in Vita, this mobile SOA framework also works as a bridge between the mobile platform and cloud platform of Vita over the Internet.

The design of the **cloud platform** of Vita is based on RESTful Web Service-based architecture design methodologies and specifications, so that an open, extensible and seamless architecture can be provided across the mobile and cloud platform of Vita. The cloud platform of Vita includes four components:

- Management interface: It provides the development environment and application programming interfaces to support application developers and enable third-party service providers to participate in the development of different applications and services for mobile crowdsensing.

- Storage service: It supports automatic backup of Vita system data, such as data related to software services, installation files of Vita in the mobile devices, task lists and results of mobile crowdsensing, and sensing data uploaded by the mobile devices through Vita.

- Deployment environment: It enables dynamic deployments of the mobile platform and various web services of Vita to different mobile devices, according to their capacities and practical application requirements.

- Process runtime environment: It provides the services of the crowdsensing platform, such as coordinating, processing, and combining multiple crowdsensing results from different mobile devices in real-time.

Vita uses a resource optimization mechanism called an **application-oriented service collaboration model** (ASCM) for allocating human-based tasks among individuals, and computing tasks between mobile devices and cloud computing platforms, efficiently and effectively. The ASCM includes three components:

- Transformer: The Transformer can transfer the diverse application requests and tasks from users to a standard data format as a web service during a mobile device's runtime.

- Task Client: The Task Client works as a bridge between the ASCM and the mobile SOA framework. It could receive the application requests of users and invokes the existing services in the mobile SOA framework to accomplish the tasks.

- Social Vector: The Social Vector can obtain application tasks information from the Transformer, and service and social information from the mobile SOA framework through the Task Client. Based on these, the Social Vector quantifies the distance and relationship between two physical elements (i.e., people, mobile devices, server of the cloud platform) and virtual elements (i.e., software service, computing/human-based task) to facilitate the deployment of computing tasks and human-based tasks among different devices and users of mobile CPS, according to different specific crowdsensing application scenarios.

5.4.2 Application Example of Vita

An application named "Smart City" was developed on Vita, to demonstrate the functionalities of Vita and the applications of mobile CPS for crowdsensing in our daily lives. *Smart City* consists of two generic functions (Services and Crowdsensing) and two application-specific functions (eating and shopping tour).

The generic functions are:

1. Services: This RESTful-based service function can be extended by application developers flexibly to enable new application-specific service-sharing strategies, which allow the users to easily share functions in mobile devices' run-time during some specific crowdsensing scenarios, via the cloud platform of Vita.

2. Crowdsensing: This function allows mobile users to post crowdsensing requests through social networks, to find out the potential people who could help to accomplish the tasks, and to accept new crowdsensing tasks by choosing the preferable task on the list.

The application-specific functions are:

Eating: The eating function consists of four sub-functions: i-Ask, Search, Comment, and Photo. It is designed to enable people who travel in a new city to conveniently find out and share food information that interests them in real-time. In the prototype, the application scenario is that a visitor called Blair has traveled from Vancouver to Hong Kong, and she wants to taste some local food in Hong Kong that Vancouver does not have. Three other persons are in restaurants. First, Blair posts the related crowdsensing request through the eating feature of the Smart City application on her phone, as follows: What food can be found here that is not available in Vancouver? Second, based on the location service, and with the help of ASCM and the cloud platform of Vita, the request can be automatically pushed to the three persons who are dining in restaurants and have experience about this (i.e., two of them had lived in Vancouver before). Third, through the user interface of Smart City, they take a photograph of the food and attach simple comments. The photograph and comments are automatically combined with the location service and map service, and then the answers are uploaded to the cloud platform of Vita. After the cloud platform of Vita gets all of the answers, it can automatically match the other answers stored in it before. The overall answers are then returned to Blair's phone. Experimental evaluation shows that the time taken by Vita to return the answers after Blair made the request was approximately 2–3 minutes.

Shopping tour: This function is designed to assist users in a city to easily find out and/or share the shopping information that interests them in real-time. This shopping example demonstrates that Vita can leverage the advantages of Internet services (i.e., today's shopping information) and cloud computing platform to enable people to aggregate and realize the shopping information that they are interested in, through mobile devices in a convenient and efficient manner.

5.4.3 Performance and Reliability

The system and task performance of Vita is evaluated in terms of three parameters: time efficiency, energy consumption, and networking overhead in mobile devices. These parameters are measured when the mobile users finish crowdsensing and concurrent computation tasks, as these parameters have great impact on the experience of mobile users when they are participating in mobile crowdsensing.

The time delay refers to the periods between the time that the smartphone initiates a crowdsensing request to the cloud platform and the time that it receives the responses from the servers on the cloud platform of Vita. The average time delay observed in the experiments using Vita's system is 11 seconds. This time delay is very low compared to Medusa's average time delay, which is about 64 seconds.

The Vita system uses a service state synchronization mechanism (S3M) [125] to detect and recover the possible service failures of mobile devices when they are running tasks and collaborating with the cloud platform of Vita. Service failures could be detected and recovered with the help of S3M, since S3M includes the function to store the stage execution state on both the mobile device and the cloud platform of Vita. The experimental results show that the average time delay with S3M loaded is 12.8 seconds, whereas the increases in battery consumption and network overhead are relatively higher, at about 75% and 43%, respectively. Based on these results, developers can choose whether or not to integrate the S3M model according to their specific purposes when developing mobile crowdsensing applications on Vita.

5.5   Other Systems

PEIR [124] and SoundSense [108] are similar participatory sensing systems, but which do not incorporate incentives into sensing. Moreover, they do not support programmable data collection from multiple sensors or an in-network processing library. The PEIR application uses mobile phones to evaluate if users have been exposed to airborne pollution, enables data sharing to encourage community participation, and estimates the impact of individual user/community behaviors on the surrounding environment. mCrowd [179] is another system that enables users to post and work on sensor-related tasks, such as requesting photos of a specific location, asking to tag photos, and monitoring traffic.

AnonySense [148] is a privacy-aware tasking system for sensor data collection and in-network processing, whereas PRISM [59] proposes a procedural programming language for collecting sensor data from a large number of mobile phones. More recently, Ravindranath et al. [140] have explored tasking smartphones crowds, and provide complex data processing primitives and profile-based compile-time partitioning.

ParticipAct [48] is another sensing platform that focuses on the management of large-scale crowdsensing campaigns as real-world experiments. The

ParticipAct platform and its ParticipAct living lab [49] form an ongoing experiment at the University of Bologna. This involves 170 students for one year in several crowdsensing campaigns that can passively access smartphone sensors and also prompt for active user collaboration.

Cardone et al. [48] proposed an innovative geo-social model to profile users along different variables, such as time, location, social interaction, service usage, and human activities. Their model also provides a matching algorithm to autonomously choose people to involve in sensing tasks and to quantify the performance of their sensing. The core idea is to build time-variant resource maps that could be used as a starting point for the design of crowdsensing ParticipActions. In addition, it studies and benchmarks different matching algorithms aimed to find, according to specific urban crowdsensing goals geolocalized in the Smart City, the "best" set of people to include in the collective ParticipAction. The technical challenge here is to find, for the specific geo-socially modeled region, the good dimensioning of number/profiles of the involved people and sensing accuracy.

Tuncay et al. [165] propose a participant recruitment and data collection framework for opportunistic sensing, in which the participant recruitment and data collection objectives are achieved in a fully distributed fashion and operate in DTN (Delay Tolerant Network) mode. The framework adopts a new approach to match mobility profiles of users to the coverage of the sensing mission. Furthermore, it analyzes several distributed approaches for both participant recruitment and data collection objectives through extensive trace-based simulations, including epidemic routing, spray and wait, profile-cast, and opportunistic geocast. The performances of these protocols are compared using realistic mobility traces from wireless LANs, various mission coverage patterns, and sink mobility profiles. The results show that the performances of the considered protocols vary, depending on the particular scenario, and suggest guidelines for future development of distributed opportunistic sensing systems.

5.6 Conclusion

In this chapter, we discussed existing mobile crowdsystems platforms. We first described the McSense system, which is a mobile crowdsensing platform that allows clients to collect many types of sensing data from users' smartphones. We then presented Medusa, a programming framework for crowdsensing that provides support for humans-in-the-loop to trigger sensing actions or review results, recognizes the need for participant incentives, and addresses their privacy and security concerns. Finally, we discussed Vita, a mobile cyber-physical system for crowdsensing applications, which enables mobile users to perform mobile crowdsensing tasks in an efficient manner through mobile devices.

6

General Design Principles and Example of Prototype

6.1 Introduction

We believe that mobile crowdsensing with appropriate incentives and with a secure architecture can achieve real-world, large-scale, dependable, and privacy-abiding people-centric sensing. However, we are aware that many challenges have to be overcome to make the vision a reality. In this chapter, we describe a general architecture for a mobile crowdsensing system considering the existing architectures presented previously in Chapter 5. We then point out important principles that should be followed when implementing the MCS prototypes in order to address known challenges in data collection, resource allocation, and energy conservation.

6.2 General Architecture

Mobile crowdsensing requires that sensing and sharing of data occurs in a robust and efficient fashion. Smartphones cannot be overloaded with continuous sensing commitments that undermine the performance of the phone (e.g., by depleting battery power). Therefore, one challenge is to determine which architectural components should run on the phone and which should run in the cloud. Some MCS architecture models propose that raw sensor data should not be pushed to the cloud because of privacy issues.

In the early phases of mobile crowdsensing development, the sensing applications were built specific to each application's requirements. Such an application-specific architecture hinders the development and deployment of MCS applications in many ways. First, it is difficult for developers to program an application. To write a new application, the developer has to address challenges in energy, privacy, and data quality. These issues are common for any MCS application, so for every developer it is like reinventing the wheel all the time. Furthermore, developers may need to develop different variants of local analytics if they want to run the application on a wide variety of devices running on different operating systems.

Second, this approach is inefficient. Applications performing sensing and processing activities independently without understanding each other's high-level context will result in low efficiency when these applications start sensing similar data from the resource-constrained smartphones. Moreover, there is no collaboration or coordination across devices, so all the devices may not be needed when the device population is dense. Therefore, the current architecture is not scalable. Only a small number of applications can be accommodated on each device. Also, the data gathered from large crowds of the public may overwhelm the network and the back-end server capacities, thus making the current architecture non-scalable.

6.2.1   Current Architecture

Currently, a typical mobile crowdsensing (MCS) system has two application-specific components, one on the device (for sensor data collection and propagation) and the second in the back-end (web server or cloud) for the analysis of the sensor data required for the MCS application. Such unifying architecture addresses the limitations discussed in application-specific architectures. It will satisfy the common needs for multiple different applications. This general architectural viewpoint for the mobile phone and the computing cloud is described as a means to discuss the major architectural issues that need to be addressed. This may not be the best system architecture, but it gives a good baseline for discussions that will eventually lead to a converging view and move the field forward.

In the general architecture, individual smartphones collect raw sensor data from the sensors embedded in the phone. Machine learning and data mining techniques can be applied to extract the information from the sensor data. These aggregation tasks execute either directly on the phone, in the back-end server or cloud, or with some partitioning between the smartphone and cloud. Decisions on where these components should execute can depend on various architectural considerations, such as privacy, providing user real-time feedback, reducing communication cost between the phone and cloud, available computing resources, and sensor fusion requirements. Hence, the issue of where these components should run remains an open problem that needs to be explored more.

Based on the commonality or coupling of the components, we can group a number of important architectural components together. For instance, a personal sensing application will only inform the user, whereas a group or community sensing application may share an aggregate version of the data with a broader population and can mask the identity of the individual contributing users.

6.2.2 Implementation Principles

The general principles that should be followed by the implementation of mobile crowd sensing systems are:

1. It should allow application developers to specify their data needs in a high-level language.

2. It should identify common data needs across applications to avoid duplicate sensing and processing activities on devices.

3. It should automatically identify the set of devices that can provide the desired data, and produce instructions to configure the sensing activities on devices properly.

4. When dynamic changes happen, it should adapt the set of chosen devices and sensing instructions to ensure the desired data quality.

5. To avoid implementing different versions of local analytics on heterogeneous devices, it is necessary to have a layer that hides the differences in physical sensor access APIs and provide the same API upwards. This makes it possible to reuse the same local analytics across different device platforms.

Other MCS architectural designs are possible. For example, it might be possible to design a purely peer-to-peer crowdsensing system, and it might also be possible to empower smartphones to exclusively execute crowdsensing tasks. The described architectural principles enable us to achieve the crowdsensing requirements without significantly complicating the system design. Furthermore, these principles do not precisely determine what functionality to place on the cloud or on the phone. In the next section, we present a prototype whose implementation adheres to most of these design principles.

6.3 Prototype Implementation and Observations

In this section, we discuss the McSense prototype implemented by Talasila et al. [19] to run a mobile crowdsensing user study. The McSense application, shown in Figure 6.1, has been implemented in Android and is compatible with smartphones running Android OS 2.2 or higher. The application was tested successfully using Motorola Droid 2 phones that have 512 MB RAM, 1 GHz processor, Bluetooth 2.1, Wi-Fi 802.11 b/g/n, 8 GB on-board storage, and 8 GB microSD storage. The McSense [19] Android application was deployed to Google Play [12] to make it available for students on campus. The server side of McSense is implemented in Java/J2EE using the MVC (Model View Controller) framework. The Derby database is used to store the registered user

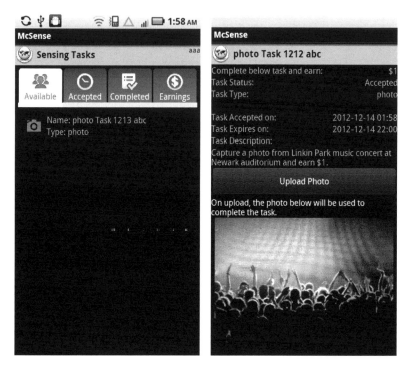

FIGURE 6.1
McSense Android Application showing tabs (left) and task screen for a photo task (right).

accounts and assigned task details. The server side Java code is deployed on the Glassfish Application Server, which is an open-source application server.

6.3.1   Task Assignment Policies

McSense deals with a continuously changing landscape: tasks are dynamically described and assigned, while participants roam free, following paths that unpredictably change dynamically, making it difficult to predict the best task-participant assignment schema. McSense evaluates the sensing task performance in terms of three main goal parameters: success ratio (i.e., ratio of successfully completed tasks over created ones), completion time (i.e., the interval between task start time and its successful conclusion), and number of required participants (i.e., number of participants to activate for task execution). At sensing task creation time, the city manager in a Smart City can either rely on default goal parameter values proposed by McSense or configure them depending on the experience stemming from previous task execution runs and specific application requirements.

For instance, for a mission-critical sensing task, she will choose higher completion probability with a higher number of participants, whereas a Smart City municipality with a low budget may be interested in minimizing the number of participants to reduce the associated incentive costs. Let us stress that these parameters also represent variables that are mutually dependent and deeply related to the assignment algorithm used to select participants.

McSense provides three different policies that, taking as inputs task properties (e.g., location, area, and duration) and user/region profiles, assign tasks to participants: random policy, attendance policy, and recency policy. The random policy is not context aware and selects the set of participants to employ as a random group of available people in the whole city. The attendance policy exploits knowledge about the time previously spent by people in the task area; based on that indicator, it chooses and ranks potentially good participants. The recency policy, instead, favors and selects as participants the people who have more recently (with respect to the creation time of the sensing task) traversed the sensing task area.

For each policy, McSense calculates ranked lists of candidate participants. In addition, all the implemented policies do not consider participants whose battery level is below a certain threshold, called battery threshold, at task starting time, because it is assumes that these participants will be unlikely to run the sensing task in order to avoid battery exhaustion. The last input parameter, called participants ratio, is used to decide the percentage of candidate participants that will receive the task assignment, expressed as a percentage value in the [0.0, 1.0] range.

Given the above inputs and assignment policies, the McSense task assignment component evaluates sensing task performance (success ratio, completion time, and number of required participants). Moreover, because completed tasks are typically much fewer than all possible crowdsensing tasks of potential interest (e.g., some of them could relate to scarcely traversed areas), the task assignment component also runs prediction algorithms to decide how to effectively self-manage its behavior based on the sensing and profiling data harvested so far. The primary goal is to exploit the already collected sensing and profiling information to reasonably calculate accurate performance forecasts and estimations about potential future tasks in a relatively lightweight way.

The prediction process consists of two phases and exploits the collected real-world dataset by dividing it temporally into two parts. The first phase considers the first part of the dataset as the past history and builds user/region profiles, such as ranked candidate participant lists. The second phase, instead, as better detailed in the following, creates synthetic ("virtual") future tasks and assigns them to candidate participants selected according to the prediction algorithms; then it evaluates the performance of the predicted situations. Task designers take advantage of those forecasts, evaluated offline by the prediction process, to get fast feedback online about the expected performance of the sensing tasks they are defining.

With a closer view to technical details, McSense task prediction stochastically generates virtual tasks with different locations, areas, and durations. It emulates their execution based on the user profile stored in the data backend, in particular, based on location traces: a virtual task is considered to be successfully completed if the location trace of a user comes to in its range. To make the model more realistic, Talasila et al also assume that participants whose device battery level is very low (e.g., less than 20 percent) will never execute any task, while if the battery level is high (e.g., 80 percent or more), they will always execute any task they can; the probability of executing a task increases linearly between 20 and 80 percent.

In particular, given a task, its duration, and the set of participants to whom it has been assigned, the emulator looks for a participant within the task area by iterating participant position records in the task duration period. When it finds one, it stochastically evaluates whether the participant will be able to complete the task, and then updates the statistics about the policy under prediction by moving to the next participant location record. In the future, additional user profile parameters can be added, such as task completion rate or quality of data provided. The predicted situations are run for each assignment policy implemented in McSense; city managers can exploit these additional data to compare possible assignment policies and to choose the one that better suits their needs, whether it has a high chance of successful completion, minimizes the completion time, or minimizes the number of participants involved.

6.3.2   User Survey Results and Observations

At the end of the field study, Talasila et al. requested each user to fill out a survey in order to understand the participants' opinion on location privacy and usage of phone resources. The survey contained 16 questions with answers on a five-point Likert scale (1="Strongly disagree," 2="Disagree," 3="Neutral," 4="Agree," 5="Strongly agree"). Out of 58 participants, 27 filled in the survey. Based on the survey answers, several interesting observations are provided, which are directly or indirectly relevant in the context of data reliability:

- One of the survey questions was: "I tried to fool the system by providing photos from other locations than those specified in the tasks (the answer does not influence the payment)." By analyzing the responses for this specific question, it can be seen that only 23.5% of the malicious users admitted that they submitted the fake photos (4 admitted out of 17 malicious). This shows that the problem stated in the article on data reliability is real and it is important to validate the sensed data;

- One survey question related to user privacy was: "I was concerned about my privacy while participating in the user study." The survey results show that 78% of the users are not concerned about their privacy. This shows that many participants are willing to trade off their location privacy for paid tasks. The survey results are correlated with the collected McSense data points. The authors posted a few sensing tasks during weekends, which is considered to be private time for the participants, who are mostly not on the campus at that time. Talasila et al. observed that 33% of the participants participated in the sensing and photo tasks, even when spending their personal time on the weekends. The authors conclude that the task price plays a crucial role (trading the user privacy) to collect quality sensing data from any location and time;

- Another two survey questions are related to the usage of phone resources (e.g., battery) by sensing tasks: 1) "Executing these tasks did not consume too much battery power (I did not need to re-charge the phone more often than once a day);" 2) "I stopped the automatic tasks (resulting in incomplete tasks) when my battery was low." The responses to these questions are interesting. Most of the participants reported that they were carrying chargers to charge their phone battery as required while running the sensing tasks and were always keeping their phone ready to accept more sensing tasks. This provides evidence that phone resources, such as batteries, are not a big concern for continuously collecting sensing data from different users and locations. Next the battery consumption measurements in detail.

6.3.3 Battery Consumption

Talasila et al. tried to determine the amount of energy consumed by the user's phone battery for collecting sensing data. Basically, the aggregation protocols are executed on the server side over the collected data. But the collected data such as Bluetooth scans at each location are crucial to infer the context of the sensed data. Therefore, the authors provide measurements for the extra battery usage caused by keeping Bluetooth/Wi-Fi radios ON to perform sensing tasks. The authors measured the readings using "Motorola Droid 2" smartphones running Android OS 2.2:

- With Bluetooth and Wi-Fi radios ON, the battery life of the "Droid 2" phone is over 2 days (2 days and 11 hours);

- With Bluetooth OFF and Wi-Fi radio ON, the battery life of the "Droid 2" phone is over 3 days (3 days and 15 hours);

- For every Bluetooth discovery the energy consumed is 5.428 Joules. The total capacity of the "Droid 2" phone battery is 18.5KJ. Hence, over 3000 Bluetooth discoveries can be collected from different locations using a fully charged phone.

6.4 Resource Management

The extensive use of sensors in smartphones by a variety of crowdsensing applications poses a significant strain on the battery life. In addition to the smartphone's battery, the communication network usage, storage space and processing capability are other smartphone resources that are also used for aggregating and transferring sensed data. To maintain the required number of participants at all times in a mobile crowdsensing system, it is important to efficiently and fairly orchestrate the use of available resources on the participants' smartphones. This involves designing the resource management layer that enables the efficient scheduling and distribution of sensor processing tasks and optimizes the utilization of resources to meet the sensing requirements. A general and simple approach that minimizes the energy consumption and network usage while collecting sensed data is to transmit data from participant smartphone only when connected to WiFi and when the phone is plugged in for charging.

The problem of incentive mechanism design for data contributors for participatory sensing applications relates indirectly to resource management. For example, in an auction-based system, the service provider receives service queries in an area from service requesters and initiates an auction for user participation. Upon request, each user reports its perceived cost per unit of amount of participation, which essentially maps to a requested amount of compensation for participation.

The participation cost quantifies the dissatisfaction caused to users due to participation. This cost may be considered to be private information for each device, as it strongly depends on various factors inherent to it, such as the energy cost for sensing, data processing and transmission to the closest point of wireless access, the residual battery level, the number of concurrent jobs at the device processor, the required bandwidth to transmit data, and the related charges of the mobile network operator, or even the user discomfort due to manual effort to submit data. Hence, participants have strong motives to misreport their cost, i.e., declare a higher cost than the actual one, so as to obtain higher payment.

Therefore, a mechanism for user participation level determination and payment allocation that is most viable for the provider is needed. It should minimize the total cost of compensating participants, while delivering a certain quality of experience to service requesters.

The mobile crowdsensing system should treat its participants as overall resources. Thus, if the tasks are allocated efficiently, then the overall redundancy in sensing tasks is reduced. In turn, this reduces the strain on each individual's smartphone resources. Such an energy-efficient allocation framework [187] was designed, whose objective is characterized by a min–max aggregate sensing time. Since optimizing the min–max aggregate sensing time is

NP hard, the authors have proposed two other allocation models: offline allocation and online allocation. The offline allocation model relies on an efficient approximation algorithm that has an approximation ratio of $2 - \frac{1}{m}$, where m is the number of participating smartphones in the system. The online allocation model relies on a greedy online algorithm, which achieves a competitive ratio of at most m. Simulation results show that these models achieve high energy efficiency for the participants' smartphones: The approximation algorithm reduces the total sensing time by more than 81% when compared to a baseline that uses a random allocation algorithm, whereas the greedy online algorithm reduces the total sensing time by more than 73% compared to the baseline.

The energy-efficient allocation framework [187] mainly focuses on the collection of location-dependent data in a centralized fashion and without any time constraints. However, there are scenarios where the service provider aims to collect time-sensitive and location-dependent information for its customers through distributed decisions of mobile users. In that case, the mobile crowdsensing system needs to balance the resources in terms of rewards and movement costs of the participants for completing tasks. Cheung et al. [53] proposed a solution to such a distributed time-sensitive and location-dependent task-selection problem. They leveraged the interactions among users as a non-cooperative task-selection game, and designed an asynchronous and distributed task-selection algorithm for each user to compute her task selection and mobility plan. Each user only requires limited information on the aggregate task choices of all users, which is publicly available in many of today's crowdsourcing platforms.

Furthermore, the primary bottleneck in crowdsensing systems is the high burden placed on the participant who must manually collect sensor data to simple queries (e.g., photo crowdsourcing applications, such as grassroots journalism [74], photo tourism [151], and even disaster recovery and emergency management [107]). The Compressive CrowdSensing (CCS) framework [177] was designed to lower such user burden in mobile crowdsourcing systems. CCS enables each participant to provide significantly reduced amounts of manually collected data, while still maintaining acceptable levels of overall accuracy for the target crowd-based system. Compressive sensing is an efficient technique of sampling data with an underlying sparse structure. For data that can be sparsely represented, compressive sensing shows the possibility to sample at a rate much lower than the Nyquist sampling rate, and then to still accurately reconstruct signals via a linear projection in a specific subspace. For example, when applied to citywide traffic speeds that have been demonstrated to have a sparse structure, compressive sensing can reconstruct a dense grid of traffic speeds from a relatively small vector that roughly approximates the traffic speeds taken at key road intersections. Naive applications of compressive sensing do not work well for common types of crowdsourcing data (e.g., user survey responses) because the necessary correlations that are exploited by a sparsifying base are hidden and non-trivial to identify. CCS comprises a series of novel techniques that enable such challenges to be overcome. Central to the CCS design is the Data Structure Conversion technique that is able to search a variety of representations of the data in an effort to find one that is then suitable for learning a custom sparsifying base (for example, to mine temporal and spatial relationships). By evaluating CCS with four representative large-scale datasets, the authors find that CCS is able to successfully lower the quantity of user data needed by crowd systems, thus reducing the burden on participants' smartphone resources. Likewise, SmartPhoto [174] is another framework that uses a resource-aware crowdsourcing approach for image sensing with smartphones.

CrowdTasker [176, 185] is another task allocation framework for mobile crowdsensing systems. CrowdTasker operates on top of the energy-efficient Piggyback Crowdsensing (PCS) task model, and aims to maximize the coverage quality of the sensing task. In addition, it also satisfies the incentive budget constraints. In order to achieve this goal, CrowdTasker first predicts the call and mobility of mobile users based on their historical records. With a flexible incentive model and the prediction results, CrowdTasker then selects a set of users in each sensing cycle for PCS task participation, so that the resulting solution achieves near maximal coverage quality without exceeding the incentive budget.

6.5 Conclusion

In this chapter, we discussed the general design and implementation principles for prototypes of mobile crowdsensing. We first presented a general architecture based on the current systems. Subsequently, we discussed the general implementation principles that are needed to build a robust mobile crowdsensing system. Finally, we presented implementation details for a mobile crowdsensing system prototype, observations from a user study based on this prototype, and mechanisms for resource management in mobile crowdsensing systems.

7

Incentive Mechanisms for Participants

7.1 Introduction

A major challenge for broader adoption of the mobile crowdsensing systems is how to incentivize people to collect and share sensor data. Many of the proposed mobile crowdsensing systems provide monetary incentives to smartphone users to collect sensing data. There are solutions based on micropayments [141] in which small tasks are matched with small payments. Social incentive techniques such as sharing meaningful aggregated sensing information back to participants were also explored, to motivate individuals to participate in sensing. In addition, there are gamification techniques proposed for crowdsourced applications [115, 81]. In this chapter, we discuss in detail each of these incentive techniques.

7.2 Social Incentives

Social incentives can be leveraged to incentivize participants in crowdsensing systems. For example, useful personal analytics are provided as incentives to participants through sharing bicycle ride details in Biketastic [143]. Biketastic is a platform that makes the route sharing process both easier and more effective. Using a smartphone application and online map visualization, bikers are able to document and share routes, ride statistics, sensed information to infer route roughness and noisiness, and media that documents ride experience. Biketastic was designed to ensure the link between information gathering, visualization, and bicycling practices.

Biketastic: The system aims to facilitate knowledge exchange among bikers by creating a platform where participants can share routes and biking experience. Biketastic participants ride new or familiar routes while running an application on a mobile phone. The mobile phone application captures location, sensing data, and media. The data collection process and the sensors involved in Biketastic are:

1. Basic route information, such as the spatial and temporal extent as well as length and speed information, is obtained by recording a location trace using the GPS sensor.

2. The physical dynamics of the route are documented using the accelerometer and microphone. Specifically, road roughness and general noise level along a route is inferred using these sensors.

3. The experience of a route is obtained by having individuals capture geo-tagged media, such as images and video clips of interesting, troublesome, and beneficial assets, along with tags and descriptions while riding.

4. The route data are uploaded to a backend platform that contains a map-based system that makes visualizing and sharing the route information easy and convenient.

LiveCompare: Another variety of social incentive is enabling data bartering to obtain additional information, such as bargain hunting through price queries in LiveCompare [63]. LiveCompare is a system that leverages the ubiquity of mobile camera phones to allow for grocery bargain hunting through participatory sensing. It utilizes two-dimensional barcode decoding to automatically identify grocery products, as well as localization techniques to automatically pinpoint store locations. The incentive scheme is inherently ingrained into the query/response protocol, and self-regulating mechanisms are suggested for preserving data integrity. At a high level, LiveCompare employs the following steps:

1. The system participants use their camera phones to snap a photograph of the price tag of their product of interest.

2. The product is uniquely identified via a barcode included on the price tag in most grocery stores.

3. The photograph is then uploaded to a central repository for satisfying future queries.

4. In exchange for submitting this price data point, the user receives pricing information for the scanned product at other nearby grocery stores.

Users only benefit from the application when the server compares their submitted price to other related data points. Therefore, the data pool can only be queried as users simultaneously contribute. LiveCompare relies on humans, rather than machines, to interpret complex sale and pricing information. The only part of a price tag that must be interpreted by a computer is the barcode to retrieve a Universal Product Code (UPC). Because of this, LiveCompare does not need to rely on error-prone OCR algorithms to extract textual tokens or on linguistic models to make sense of sets of tokens. Furthermore, each LiveCompare query returns a subset of the data pool for a user to consider. If an image does not seem relevant, the user can quickly flag it. This allows users to collectively identify faulty or malicious data.

7.3 Monetary Incentives

Monetary incentives are the most commonly used incentives in mobile crowd-sensing systems. McSense [50] is one such system that uses monetary incentives. The architecture and design of McSense is detailed in Section 5.2 and its implementation in Section 6.3. McSense is a micro-payment-based system that allows the participants to choose from a wide range of sensing tasks, such as taking photos at events on campus, collecting GPS and accelerometer readings, or collecting application and network usage from the phones. When choosing a task, the participants have to balance the value of micro-payments (different for each task) against their effort, the potential loss in privacy, and the resource consumption on the phone (e.g., battery). A McSense application was implemented for Android phones.

Talasila et al. [159, 157] evaluated the effectiveness of micro-payments in crowdsensing through a user study that sought to understand: (1) the efficiency of area coverage, (2) the reliability of user-provided data, and (3) the relation between monetary incentives and task completion.

This study ran for 2 months with 50 students, who volunteered and registered at the beginning of the study. The photo tasks that require more effort from participants are valued at $10, $5, and $3 per task. The photo tasks that require less effort by participants are valued at $2, $1, and 50 cents per task. The automated sensing tasks require less effort by participants and are worth between 50 cents and $2 per task.

Area coverage in mobile crowdsensing systems using micro-payments. Talasila et al. investigated the impact on area coverage based on the collected WiFi data while using micro-payments as incentive. These data were collected as part of the *resource usage task* in the last 28 days of the McSense study. During this study, there were always resource usage tasks available to users. Furthermore, the users were directed to various regions on campus by other tasks (e.g., taking photos of certain events), thus improving the sensing coverage. Specifically, this method allows for capturing additional indoor WiFi data from places where users do not typically go. For example, the participants were asked to capture photos of recycle bins at each floor in the campus buildings along with room numbers. The system administrator made sure there were always photo tasks available for the participants on each day of the user study (including weekends).

Figure 7.1 shows the area coverage efficiency of the micro-payments-based approach. Based on the WiFi signal strength, there are areas with strong, medium, low, or no signal. There is a relatively constant increase in the cover-

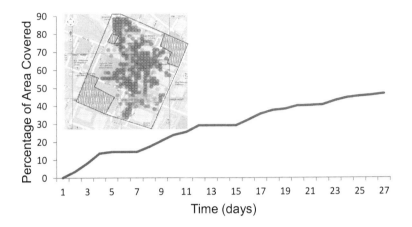

FIGURE 7.1
Area coverage over time while performing crowdsensing with micro-payments.
Illustration of the coverage overlaid on the campus map. Some areas have
been removed from the map as they are not accessible to students.

age over time in the first 3 weeks, especially during weekdays. Toward the end
of the study, the rate of coverage decreases as most of the common areas have
been covered. The authors speculate that users did not find the price of the
tasks enticing enough to go to areas located far away from their daily routine.
Overall, the micro-payments-based approach achieves a 46% area coverage of
the campus over the four-week time period of the study.

Other monetary incentive schemes: Another solution based on monetary incentives [96] incorporates a metric called users' quality of information
(QoI) into its incentive mechanisms for mobile crowdsensing systems (MCS).
Due to various factors (e.g., sensor quality, noise, etc.), the quality of the
sensory data contributed by individual users varies significantly. It is always
desirable to obtain high-quality data while minimizing the expenses in MCS
platforms. This solution is based on reverse combinatorial auctions, in which
the platform acts as the auctioneer that purchases the data from participating users. Two models are investigated: the single-minded model in which a
user is willing to execute one subset of tasks, and the multi-minded model
in which a user might be interested in executing multiple subsets of tasks.
For the former model, the authors design a truthful, individual, rational,
and computationally efficient incentive mechanism that maximizes the social welfare within a guaranteed approximation ratio. For the latter model,
the authors design an iterative descending mechanism that achieves close-
to-optimal social welfare within a guaranteed approximation ratio, while
satisfying individual rationality and computational efficiency.

In addition, the design of quality-based incentive mechanisms is necessary to avoid inefficient sensing and unnecessary rewards. By considering the data quality in the design of incentive mechanisms for crowdsensing, Peng et al. [130] propose a monetary incentive solution that pays the participants depending on how well they perform. This motivates rational participants to perform data sensing correctly and efficiently. This mechanism estimates the quality of sensing data and offers participants a reward based on their effective contribution. The evaluation results of this scheme show that the mechanism achieves superior performance in terms of quality assurance and profit management when compared to the uniform pricing scheme, which regards all sensed data equally and as being of the same quality, and pays equal rewards to participants.

7.4 Mobile Game Incentives

Gaming is another incentive technique that has high potential to attract participants in mobile crowdsensing. While there is a significant literature on using gamification techniques in crowdsourcing, there is very little in terms of applying gamification techniques in mobile sensing. Gaming could be a cost-effective alternative to micro-payments when attracting users to participate in MCS systems. For example, "Alien vs. Mobile User" [2] is a first-person shooter sensing game that is played by mobile crowdsensing participants on their smartphones. The game involves tracking the location of extraterrestrial aliens on the campus map of a university and destroying them. The game entices users to unpopular regions through a combination of in-game incentives, which include alien-finding hints and a higher number of points received for destroying aliens in these regions. The game was implemented in Android, and it collects WiFi signal data to construct the WiFi coverage map of the targeted area. In next sub-sections, we provide the design, implementation, and user study details of the mobile game incentive technique proposed by Talasila et al. [2].

7.4.1 Alien vs. Mobile User Game

Mobile crowdsensing enables real-time sensing of the physical world. However, providing uniform sensing coverage of an entire area may prove difficult. It is possible to collect a disproportionate amount of data from very popular regions in the area, while the unpopular regions remain uncovered. To address this problem, Talasila et al. propose a model for collecting crowdsensing data based on incentivizing smartphone users to play sensing games, which provide in-game incentives to convince participants to cover all the regions of a target area.

Since the goal of the game is to uniformly cover a large area with sensing data, it is essential to link the game story to the physical environment. In the game "Alien vs. Mobile User," the players must find aliens throughout an area and destroy them using bullets that can be collected from the target area. The players collect sensing data as they move through the area. Although the game could collect any type of sensing data available on the phones, the implementation collects WiFi data (*BSSID, SSID, frequency, signal strength*) to build a WiFi coverage map of the targeted area. The motivation to play the sensing game is twofold: 1) The game provides an exciting *real-world gaming experience* to the players, and 2) The players can *learn useful information about the environment*, such as the WiFi coverage map, which lists the locations having the best WiFi signal strength near the player's location.

7.4.2   Design and Implementation

The game contains the following entities/characters:

- *CGS*: The Central Game Server (CGS) controls the sensing game environment on the mobile devices and stores the players' profiles and the collected sensing data.

- *Player*: Users who play the game on their mobile devices.

- *Alien*: A negative-role character that needs to be found and destroyed by the players. Aliens are controlled by CGS according to the sensing coverage strategy.

Game story: The aliens in the game are hiding at different locations across the targeted area. Players can see the aliens on their screens only when they are close to the alien positions. This is done in order to encourage the players to walk around to discover aliens; in the process, the CGS collects sensing data. At the same time, this makes the game more unpredictable and potentially interesting. The game periodically scans for nearby aliens and alerts the players when aliens are detected; the player locates the alien on the game screen and starts shooting at the alien using the game buttons. When an alien gets hit, there are two possible outcomes: If this is the first or second time the alien is shot, the alien escapes to a new location to hide from the player. To completely destroy the alien, the player has to find and shoot the alien three times, while hints of the alien's location are provided after it was shot. In this way, the players are provided with an incentive to cover more locations. Players are rewarded with points for shooting the aliens. All players can see the current overall player ranking.

The sensing side of the game: Sensing data is collected periodically when the game is on. The placement of aliens on the map seeks to ensure uniform sensing coverage of the area. The challenge, thus, is how to initially place and then move the aliens to ensure fast coverage, while at the same time maintaining a high player interest in the game.

In the initial phases of sensing, CGS moves each alien to a location that is not yet covered, but later on it moves the alien intelligently from one location to another by considering a variety of factors (e.g., less visited regions, regions close to pedestrian routes, or regions that need higher sensing accuracy). In this way, the game manages to entice users from popular regions to unpopular ones with a reasonable coverage effort. Generally, the alien will escape to farther-away regions, and the players might be reluctant to follow despite the hints provided by CGS. To increase the chances that players follow the alien, the game provides more points for shooting the alien for a second time, and even more for the third (fatal) shot.

Game difficulty and achievements: The game was designed with difficulty levels based on the number of killed aliens, the bullets collected from around the player's location, and the total score of the player. In this way, players have extra-incentives to cover more ground. A player has to track and kill a minimum number of aliens to unlock specific achievements and to enter the next levels in the game. We leverage the achievements APIs provided in the Android platform as part of Google Play Game Services, which allow the players to unlock and display achievements, as shown in Figure 7.2 (right).

Prototype Implementation: A game prototype was implemented for Android-based smart-phones and was deployed on Google Play. An alien appears on the map when the player is close to the alien's location, as shown in Figure 7.2 (left). The player can target the alien and shoot it using the smartphone's touch screen. When the alien escapes to a new location, its "blood trail" to the new location is provided to the player as a hint to track it down (as shown in Figure 7.2 (middle)). The server side of the game is implemented in Java using one of the Model View Controller frameworks involving EJBs/JPA models, JSP/HTML views, and servlets, and it is deployed on the Glassfish Application Server.

7.4.3 User Study

The benefits of gamification for crowdsensing were evaluated through a user study that seeks to: (1) evaluate the area coverage efficiency, and (2) analyze the players' activity and individual contributions to area coverage.

This study ran for 35 days, during which 53 players used their Android devices to play the game and collect WiFi data outdoors and indoors. The users were not selected in any particular way. To advertise the game, the authors placed fliers throughout the campus and sent emails to students enrolled in computing majors; users continuously registered throughout the study period.

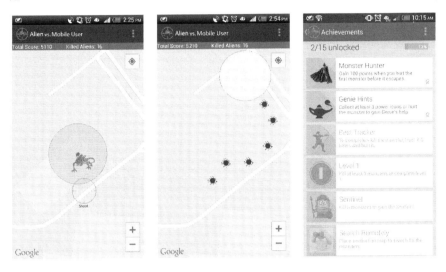

FIGURE 7.2
Alien vs. Mobile User app: finding and shooting the alien (left); alien "blood trail" (middle); player achievements (right).

Outdoor area coverage. Figure 7.3 shows the area coverage efficiency. To make the results as consistent as possible with the results from the micro-payments-based approach, the authors investigated the coverage based on the collected WiFi data from both studies (McSense and game). The authors observed that players get highly engaged in the game from the first days, which leads to high coverage quickly (50% of the target area is covered in less than 3 days). The coverage progress slows down after the initial phase due to several reasons. First, the results show only the coverage of ground level. However, starting in the second week, aliens have also been placed at higher floors in buildings; this coverage is not captured in the figure. Second, the slowdown is expected to happen after the more common areas are covered, as the players must go farther from their usual paths. Third, the authors observe that the coverage remains mostly constant over the weekends as the school has a high percentage of commuters, and thus mobile users are not on campus (as is seen on days 4 to 6, and 11 to 13).

Figure 7.3 also overlays the collected WiFi data over the campus map. The WiFi signal strength data is plotted with the same color coding as in the McSense study. Overall, the sensing game achieved 87% area coverage of the campus in a four-week period.

Indoor area coverage. Figure 7.4 plots the correlation of active players and the number of squares covered at upper floors over time (the authors started to place aliens on upper floors on day 12). Indoor localization was achieved based on WiFi triangulation and the barometric pressure sensor in

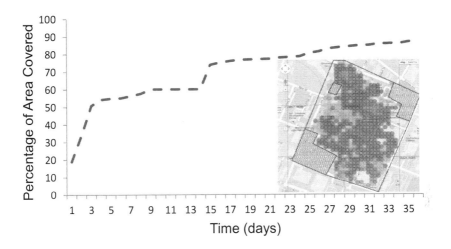

FIGURE 7.3

Area coverage over time while performing crowdsensing with mobile gaming for the first four weeks of the user study. Illustration of the coverage overlayed on the campus map. Some areas have been removed from the map as they are not accessible to students.

the phones. The authors observe that indoor coverage correlates well with the number of active players, and the pattern is similar to outdoor coverage. Overall, the game achieved a 35% coverage of the upper floors. Despite apparently being low, this result is encouraging: The players covered many hallways and open spaces in each building, but could not go into offices and other spaces that are closed to students; however, aliens were placed there as well. To avoid placing aliens in such locations and wasting players' effort, the authors plan to investigate a crowdsourcing approach in which the players mark the inaccessible places while playing the game.

Player activity. Figure 7.5 presents the impact of the number of registered players and the number of active players on area coverage over time. The results show the improvement in area coverage with the increase in the number of registered players in the game. This proves that the players are interested in the game and are involved in tracking the aliens. The players are consistently active in the weekdays over the period of the study, and they are less active in the weekends. For additional insights on the individual contribution of the players, Figure 7.6 presents the players' ranks based on number of covered squares in the area. We observe a power-law distribution of the players' contribution to the area coverage.

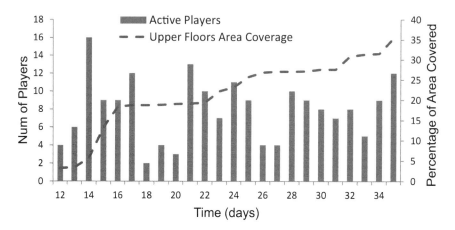

FIGURE 7.4
Correlation of active players and the number of squares covered at different floor levels over time in the last two weeks of the user study.

7.5   Comparison of Incentives

This section presents a comparison between mobile game vs monetary incentives based on the observations of both field studies.

7.5.1   Sensing Task Duration

The duration of a task plays an important role in determining whether gaming or micro-payments would be more suitable for a crowdsensing campaign. For the gaming approach to be efficient, it is extremely important that the game design includes strong in-game incentives such that the players remain engaged over time. The stronger the in-game incentives are, the longer the gaming approach remains effective. From the user studies, Talasila et al. conclude that for medium-term sensing efforts both gaming and micro-payments approaches could be effective (the two studies ran for a comparable amount of time).

However, for much longer sensing efforts (e.g., 1 year), the authors speculate that micro-payments would be more effective because monetary incentives do not depreciate, whereas in-game incentives might not be sustainable over such long periods. Similarly, the authors expect the micro-payments to perform better for tasks with tight time constraints such as taking a photo from a given event. In the study, the participants were willing to take a photo at a particular location during a particular time of the day for a reasonable micro-payment. In this regard, the gaming approach might not be as effective in quickly luring away players from their current path/activity.

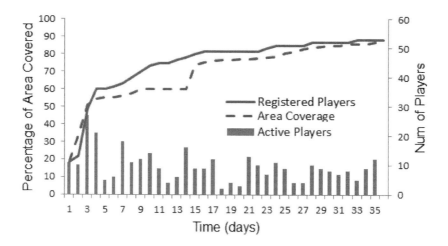

FIGURE 7.5
Impact of the number of registered players and the number of active players on area coverage over time in the user study.

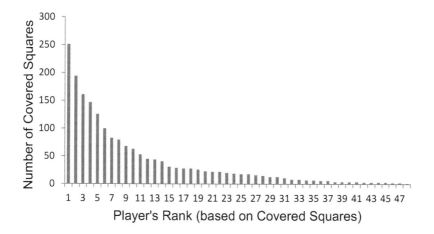

FIGURE 7.6
Ranking of players based on the number of covered squares in the area (area is divided into 10x10m squares). 48 out of 53 players covered ground on campus.

7.5.2 Sensing Task Type

It is important to note that there is no "one-size fits all" solution for all types of crowdsensing [144]. Consider the following two types of crowdsensing: 1) uniform area coverage sensing, and 2) participant activity sensing. From the observations, the gaming solution works well for area coverage sensing. However, it may not be the right fit for participant activity sensing because gaming would lead to changes in participant activities, and thus will not capture the desired sensing data of the participant. Instead, the micro-payments-based solution will be more suitable to capture the expected personal analytics.

In the mobile game study, the focus was mainly on automatically collecting sensing data, where players are not annoyed with any manual sensing tasks. In principle, micro-payments are a better fit for manual sensing. However, mobile games can also be used for this type of sensing if the sensing task does not have tight time constraints and its requirement can be translated into a game action. For example, a player could receive an in-game request to take a photo of the location where a game character was destroyed.

7.5.3   Incentive Quality or Value

The amount of collected crowdsensing data and the quality of the area coverage, where required, depend strongly on the incentive quality or value. To understand the effectiveness of in-game incentives, Talasila et al. collected game feedback from the players at the end of the study by asking them "What made you to continue playing the Alien vs. Mobile User game"? They received answers from 16 players. The responses show that the majority of the players were curious about the game story and they liked tracking the aliens hiding in the campus buildings. Other primary reasons for playing were: moving to the next game levels and being on top of the leaderboard, competing with friends, winning game achievements, and checking the game graphics. Furthermore, the authors analyzed various game variables, such as the number of players who unlocked the achievements and the mean time to unlock achievements and complete each game level, to understand the player engagement in the game. The results, omitted here, demonstrated that the game levels and achievements worked reasonably well; they were challenging enough and sustained players' interest during the study.

In the micro-payments study, the average price of the posted task was $1.18 and the average price of the completed task was $0.84. To observe the impact of price on task completion, the authors posted a few tasks with a higher price ($2 – $10). They noticed a 15% increase in task completion success rate for these tasks compared with the rest of the tasks. In addition, the authors noticed an improvement in data quality for the high-priced photo tasks, with clear and focused photos compared to the low-priced photo tasks (i.e., the tasks priced $1 or lower). Thus, the study confirms that task pricing influences the data quality. However, it is not clear whether further increase in the task price may address the issue of uniform area coverage in a cost-effective manner. On the other hand, as demonstrated by the gaming study, in-game incentives proved to be a cost-effective solution for area coverage.

7.5.4 Sensing Data Reliability

As emphasized by the results of the micro-payment study, data reliability could be a significant problem in crowdsensing. The analysis of results indicate a high correlation between the amount of time spent by users out of campus and the number of fake photos they submitted. This observation suggests that a first step toward ensuring data reliability is to incorporate location authentication mechanisms in crowdsensing. This solution applies independent of the type of incentives. In the absence of such a mechanism, statistical analysis of user's mobility traces, if available, could provide hints on which data points should be inspected critically (i.e., those from places infrequently visited by the user). Mobility traces could be more available for the gaming approach, as the games are expected to be designed to adapt to the user's location.

7.5.5 Mobile Resource Consumption

The sensing tasks should consume little resources, especially battery and cellular data, on the smartphones if crowdsensing is to be successful. For example, games should offload computationally expensive tasks to the servers/cloud. Since cellular data consumption can lead to overcharges, an option is to give players the ability to control the frequency of game status updates when using cellular data, thus choosing the desired trade-off between game accuracy and saving the phone's resources.

Similarly, long-running micro-payment tasks such as daily accelerometer and location data collection for activity recognition should be designed to ensure the phones do not run out of battery due to these tasks. The results from the micro-payment study showed that some users decided to abort the tasks when the power dropped under a certain level. However, Talasila et al. also observed that many users were recharging their phones during the day, presumably to complete the tasks and receive the monetary incentive. Existing research on continuous sensing on the phones seems promising, and once widely deployed it will alleviate such issues.

7.5.6 How General Are These Insights?

The results of the user studies provided the authors with valuable information about the effectiveness of crowdsensing based on gaming and based on micro-payments for a medium-size urban university campus: an area of 1600 x 1200 feet with buildings between 4–6 floors, and student participants mostly between 18–22 years old. One can imagine a variety of crowdsensing efforts that target similar settings, and the findings would hold in those settings. Besides university campuses, business districts or even urban neighborhoods with many young and technology-savvy professionals could achieve similar results. In addition, not every social category and age group has to be represented for certain sensing tasks such as mapping a region with sensor readings. For example, uniform area coverage is not expected to be strongly dependent on the demographics of the participants and does not require a very large number of players in the gaming approach. Our results showed that a relatively small number of passionate players quickly covered a large region.

Ideally, a wider exploration of different alternative designs of the experiments would have provided additional insights. For example, one could imagine a scenario in which each user is asked to perform data collection tasks based on micro-payments and to play the game alternatively during the study. This was not feasible given the resources available for the project. Finally, for other types of sensing tasks, such as collecting personal analytics data, the results may vary as a function of the area size as well as the population type and size.

7.6   Conclusion

In this chapter, we discussed incentive mechanisms for participants of mobile crowdsensing. We first discussed social incentives such as sharing meaningful aggregated sensing information back to participants in order to motivate individuals to participate in sensing. We then discussed monetary incentives such as micro-payments in which small tasks are matched with small payments. Subsequently, we discussed mobile game-based incentives in which participants play the sensing game on their smartphones while collecting the sensed data. Finally, we compared the incentive mechanisms in order to derive general insights.

8

Security Issues and Solutions

8.1   Introduction

This chapter examines security-related issues introduced by the mobile crowd-sensing platforms, such as ensuring the reliability, quality, and liveness of the sensed data. By leveraging smartphones, we can seamlessly collect sensing data from various groups of people at different locations using mobile crowdsensing. As the sensing tasks are associated with monetary incentives, participants may try to fool the mobile crowdsensing system to earn money. The participants may also try to provide faulty data in order to influence the outcome of the sensing task. Therefore, there is a need for mechanisms to validate the quality of the collected data efficiently. This chapter discusses these issues in detail and presents solutions to address them.

8.2   Sensed Data Reliability Issues

In the following, we motivate the need for data reliability mechanisms by presenting several scenarios involving malicious behavior.

Traffic jam alerts [85, 175]: Suppose that the Department of Transportation uses a mobile crowdsensing system to collect alerts from people driving on congested roads and then distributes the alerts to other drivers. In this way, drivers on the other roads can benefit from real-time traffic information. However, the system has to ensure the alert validity because malicious users may try to pro-actively divert the traffic on roads ahead in order to empty these roads for themselves.

Citizen journalism [163, 25]: Citizens can report real-time data in the form of photos, video, and text from public events or disaster areas. In this way, real-time information from anywhere across the globe can be shared with the public as soon as the event happens. But malicious users may try to earn easy money by claiming that the submitted data is from a certain location, while they are in reality at a different location.

Environment [26, 17]: Environment protection agencies can use pollution sensors installed in the phones to map with high accuracy the pollution zones

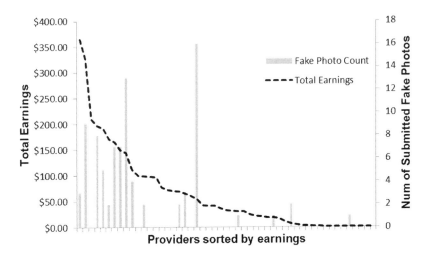

FIGURE 8.1
Correlation of earnings and fake photos.

around the country. The participants may claim "fake" pollution to hurt business competitors by claiming that the submitted sensed pollution data is associated with incorrect locations.

In Section 6.3.2, we presented observations of the "McSense" user survey that were collected from users at the end of the field study to understand the participant's opinion on location privacy and usage of phone resources . We now present insights on data reliability based on the analysis of the data collected from the "McSense" field studies

8.2.1 Correlation of User Earnings and Fake Photos

To understand the correlation between the user earnings and the number of fake photos submitted, Talasila et al. [156] plot the data collected from the McSense crowdsensing field study. The experimental results in Figure 8.1 show that the users who submitted most of the fake photos are among the top 20 high earners (with an exception of 4 low-earning users who submitted fake photos once or twice). This is an interesting observation that can be leveraged to improve the sensed data quality.

8.2.2 Correlation of Location and Fake Photos

Tallish et al. [156] ask the question "Is there any correlation between the amount of time spent by users on campus and the number of submitted fake

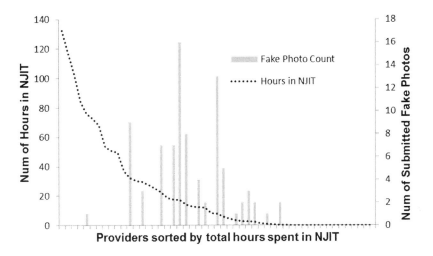

FIGURE 8.2
Correlation of user location and fake photos.

photos?" As suspected, the users who spent less time on campus have submitted more fake photos. This behavior can be observed in Figure 8.2.

Figure 8.2 shows the number of fake photos submitted by each user, with the users sorted by the total hours spent on the New Jersey Institute of Technology (NJIT) campus. The participants' total hours recorded at NJIT campus are the hours that are accumulated from the sensed data collected from "Automated Sensing task" described in the "Tasks Developed for McSense" Section 5.2.2. The NJIT campus is considered to be a circle with a radius of 0.5 miles. If the user is in this circle, then she is considered to be at NJIT. For most of the submitted fake photos with the false location claim, the users claimed that they are at a campus location where the photo task is requested, but actually they are not frequent visitors on the campus.

8.2.3   Malicious User: Menace or Nuisance?

The photos submitted by malicious users are plotted in Figure 8.3. The data show that malicious users have submitted good photos at a very high rate compared to the number of fake photos. These malicious users are among the high earners, so they are submitting more data than the average user. Thus, it may not be a good idea to remove the malicious users from the system as soon as they are caught cheating.

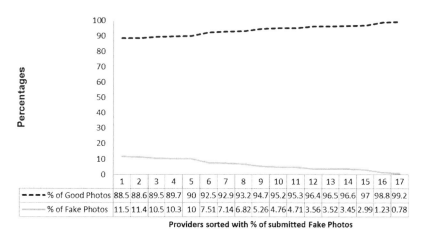

	1	2	3	4	5	6	7	8	9	10	11	12	13	14	15	16	17
----% of Good Photos	88.5	88.6	89.5	89.7	90	92.5	92.9	93.2	94.7	95.2	95.3	96.4	96.5	96.6	97	98.8	99.2
—— % of Fake Photos	11.5	11.4	10.5	10.3	10	7.51	7.14	6.82	5.26	4.76	4.71	3.56	3.52	3.45	2.99	1.23	0.78

Providers sorted with % of submitted Fake Photos

FIGURE 8.3
Photo counts of 17 cheating people.

8.3 Proposed Solutions for Data Reliability

A major challenge for broader adoption of these sensing systems is that the sensed data submitted by the participants is not always reliable [68] as they can submit false data to earn money without executing the actual task. Clients need guarantees from the mobile crowdsensing system that the collected data is valid. Hence, it is very important to validate the sensed data. However, it is challenging to validate each and every sensed data point of each participant because sensing measurements are highly dependent on context. One approach to handle the issue is to validate the location associated with the sensed data point in order to achieve a certain degree of reliability on the sensed data. Therefore, Talasila et al. [155] focus on validating the location data submitted by the participants. Still, a major challenge exists: how to validate the location of data points in a scalable and cost-effective way without help from the wireless carrier? Wireless carriers may not help with location validation for legal reasons related to user privacy or even commercial interests.

To achieve reliability of participants' location data, there are a few traditional solutions, such as using Trusted Platform Modules (TPM) [30] on smartphones or duplicating the tasks among multiple participants. However, these cannot be used directly, for a variety of reasons. For example, it is not cost-effective to have TPM modules on every smartphone, while task replication may not be feasible at some locations due to a lack of additional users there. Another solution is to verify location through the use of secure location verification mechanisms [154, 145, 46, 45] in real time when the participant is trying to submit the sensing data location. Unfortunately, this solution requires infrastructure support and adds a very high overhead on users' phones if it is applied for each sensed data point.

8.3.1 The ILR Validation Scheme over Collected Data

In this section, we present the ILR scheme [155] proposed by Talasila et al., which seeks to improve the location reliability of mobile crowdsensed data with minimal human efforts. We also describe the validation process used by McSense to detect false location claims from malicious providers.

Assumptions: Before going into the details of the scheme, the authors assume that the sensed data is already collected by the McSense system from providers at different locations. However, this sensed data is awaiting validation before being sent to the actual clients who requested this data.

For ILR, Talasila et al. assume that the sensed data includes location, time, and a Bluetooth scan performed at the task's location and time. The main idea of this scheme is to corroborate data collected from manual (photo) tasks with co-location data from Bluetooth scans. We describe next an example of how ILR uses the photo and co-location data.

Adversarial Model: Talasila et al. assume all the mobile devices are capable of determining their location using GPS. The authors also assume McSense is trusted and the communication between mobile users and McSense is secure. In the threat model, the authors consider that any provider may act maliciously and may lie about their location.

A malicious provider can program the device to spoof a GPS location [94] and start providing wrong location data for all the crowdsensing data requested by clients. Regarding this, the authors consider three threat scenarios, where 1) The provider does not submit the location and Bluetooth scan with a sensing data point; 2) The provider submits a Bluetooth scan associated with a sensing task, but claims a false location; 3) The provider submits both a false location and a fake Bluetooth scan associated with a sensing data point. In Section 8.3.1.3, we will discuss how these scenarios are addressed by ILR.

The authors do not consider colluding attack scenarios, where a malicious provider colludes with other providers to show that she is present in the Bluetooth co-location data of others. It is not practically easy for a malicious provider to employ another colluding user at each sensing location. Additionally, these colluding attacks can be reduced by increasing the minimum node degree requirement in co-location data of each provider (i.e., a provider P must be seen in at least a minimum number of other providers' Bluetooth scans at her claimed location and time). Therefore, it becomes difficult for a malicious provider to create a false high node degree by colluding with real co-located people at a given location and time.

Finally, the other class of attacks that are out of scope are attacks in which a provider is able to "fool" the sensors to create false readings (e.g., using the flame of a lighter to create the false impression of a high temperature), but submits the right location and Bluetooth scan associated with this sensing task.

8.3.1.1   An Example of ILR in Action

Figure 8.4 maps the data collected by several different tasks in McSense. The figure shows 9 photo tasks [marked as A to I] and 15 sensing tasks [marked as 1 to 15] performed by different providers at different locations. For each of these tasks, providers also report neighbors discovered through Bluetooth scans. All these tasks are grouped into small circles using co-location data found in Bluetooth scans within a time interval t. For example, Photo task A and sensing tasks (1, 2, and 3) are identified as co-located and grouped into one circle because they are discovered in each others' Bluetooth scans.

In this example, McSense does not need to validate all the photo tasks mapped in the figure. Instead, McSense will first consider the photo tasks with the highest node degree (*NodeDegree*) by examining the co-located groups for photo task providers who have seen the highest number of other providers in Bluetooth scans around them. In this example Talasila et al. consider $NodeDegree \geq 3$. Hence, we see that photo tasks A, B, C, D, and G have discovered the highest number of providers around their location. Therefore, McSense will choose these 5 photo tasks for validation. These selected photo tasks are validated either manually or automatically (we discuss this in detail in Section 8.3.1.2). When validating these photo tasks, if the photo is not valid then its photo is rejected and McSense ignores its Bluetooth scans. If the photo is valid then McSense will consider the location of the validated photo as trusted because the validated photo is actually taken from the physical location requested in the task. However, it is very difficult to categorize every photo as a valid or a fake photo. Therefore, some photos will be categorized as "unknown" when a decision cannot be made.

In this example, the authors assume that these 5 selected photos are successfully validated through manual verification. Next, using the transitivity property, McSense will extend the location trust of validated photos to other co-located providers' tasks, which are found in the Bluetooth scans of the A, B, C, D, and G photo tasks. For example, A will extend trust to the tasks 1, 2, and 3, and B will extend trust to tasks 4, 5, and 6. Now task 6 will extend its trust to tasks 13 and 14. Finally, after the end of this process, the McSense system will have 21 successfully validated tasks out of a total of 24 tasks. In this example, McSense required manual validation for just 5 photo tasks, but using the transitive property it was able to extend the trust to 16 additional tasks automatically. Only 3 tasks (E, F, and 12) are not validated as they lack co-location data around them.

8.3.1.2 The ILR Phases

The ILR scheme has two phases, as shown in Figure 8.5. "Phase 1: Photo Selection" selects the photo tasks to be validated. "Phase 2: Transitive Trust" extends the trust to data points co-located with the tasks elected in Phase 1.

Phase 1 — Photo Selection: Using collected data from Bluetooth scans of providers, the authors constructed a connected graph of co-located data points for a given location and within a time interval t (these are the same groups represented in circles as discussed in the above example). From these graphs, the authors elected the photo tasks that have node degree greater than a threshold (N_{th}).

These selected photo tasks are validated either by humans or by applying computer vision techniques. For manual validation, McSense could rely on other users recruited from Amazon MTurk [3], for example. In order to apply computer vision techniques, first we need to collect ground truth photos to train image recognition algorithms. One alternative is to have trusted people collect the ground truth photos. However, if the ground truth photos are collected through crowdsensing, then they have to be manually validated as well. Thus, reducing the number of photos that require manual validation is an important goal for both manual and automatic photo recognition. Once the validation is performed, the location of the validated photo task is now considered to be reliable because the validated photos have been verified to be taken from the physical location requested in the task. For simplicity, the authors refer to the participants who contributed valid photo tasks with reliable location and time as "Validators."

Phase 2 — Transitive Trust: In this phase, the authors rely on the transitive property and extend the trust established in the Validator's location to other co-located data points. In short, if the photo is valid, the trust is extended to co-located data points found in Bluetooth discovery of the validated photo task. In the current scheme, trust is extended until all co-located tasks are trusted or no other task is found; alternately McSense can set a TTL (Time To Live) on extended trust. The following two steps are performed in this phase:

- (Step 1) Mark co-located data points as trusted: For each task co-located with a validated photo task, mark the task's location as trusted.

- (Step 2) Repeat Step 1 for each newly validated task until all co-located tasks are trusted or no other task is found.

Validation Process: After executing the two phases of the ILR scheme, all the co-located data points are validated successfully. If any malicious provider falsely claims one of the validated tasks' location at the same time, then the false claim will be detected in the validation step. Executing the validation process shown in algorithm 1 will help us detect wrong location

---

Algorithm 1 ILR Validation Pseudo-Code

Notation:
TList: Tasks List which are not yet marked trusted after completing first two
phases of ILR scheme.
T: Task submitted by a Provider.
L: Location of the Photo or Sensing Task (T).
t: Timestamp of the Photo or Sensing Task (T).
hasValidator(L, t): Function to check, if already there exist any valid data
point at task T's location and time.

validationProcess():
run to validate the location of each task in TList
1: **for** each task T in TList **do**
2:     **if** $hasValidator(L,t) == TRUE$ **then**
3:         Update task T with false location claim at (L, t)

---

claims around the already validated location data points. For instance, if we consider task 12 from Figure 8.4 as a malicious provider claiming a false location exactly at photo task A's location and time, then task 12 will be detected in the validationProcess() function as it is not co-located in the Bluetooth scans of photo task A. In addition to the validation process, McSense will also do a basic spatio-temporal correlation check to ensure that the provider is not claiming locations at different places at the same time.

8.3.1.3   Security Analysis

The goal of the ILR scheme is to establish the reliability of the sensed data by validating the claimed location of the data points. In addition, ILR seeks to detect false claims made by malicious participants.

ILR is able to handle all three threat scenarios presented in the adversarial model. In the first threat scenario, when there is no location and Bluetooth scan submitted along with the sensed data, the sensed data of that task is rejected and the provider will not be paid by McSense.

In the second threat scenario, when a provider submits its Bluetooth discovery with a false location claim, then McSense will detect the provider in its neighbors' Bluetooth scans at a different location using the spatio-temporal correlation check, and will reject the task's data.

Finally, when a provider submits fake Bluetooth discovery with a false location claim, then the scheme looks for any validator around the claimed location, and if it finds anyone, then the sensed data associated with the false location claim is rejected. But if there is no validator around the claimed location, then the data point is categorized as "unknown."

As discussed in the adversarial model (Section 8.3.1), sensed data submitted by malicious colluding attackers could be filtered to a certain extent in McSense by setting the node degree threshold (N_{th}) to the minimum node degree requirement requested by the client.

TABLE 8.1

Photo task reliability

Total photos	1784
Number of photos with Bluetooth scans (manually validated in ILR)	204
Trusted data points added by ILR	148

TABLE 8.2

Number of false location claims

	Detected by ILR scheme	Total	Percentage detected
Tasks with false location claim	4	16	25%
Cheating people	4	10	40%

8.3.1.4 Experimental Evaluation: Field Study

The providers (students shown in Table 5.1) registered with McSense and submitted data together with their userID. Both phases of ILR and the validation process are executed on data collected from the providers, and the authors acted as the clients collecting the sensed data in these experiments.

The location data is mostly collected from the university campus (0.5 mile radius). The main goal of these experiments is to determine how efficiently the ILR scheme can help McSense to validate the location data and detect false location claims. ILR considers the Bluetooth scans found within 5min intervals of measuring the sensor readings for a sensing task.

Table 8.1 shows the total photo tasks that are submitted by people; only 204 photo tasks have Bluetooth scans associated with them. In this data set, the authors considered the $NodeDegree \geq 1$, therefore they used all these 204 photo tasks with Bluetooth scans in Phase 1 to perform manual validation, and then in Phase 2 they were able to automatically extend the trust to 148 new location data points through the transitive closure property of ILR.

To capture the ground truth, the authors manually validated all the photos collected by McSense in this study and identified that they had a total of 45 fake photos submitted to McSense from malicious providers, out of which only 16 fake photo tasks had Bluetooth scans with false location claims. Talasila et al. then applied ILR to verify how many of these 16 fake photos can be detected.

The authors were able to catch 4 users who claimed wrong locations to make money with fake photos, as shown in Table 8.2. Since the total number of malicious users involved in the 16 fake photo tasks is 10, ILR was able to detect 40% of them. Finally, ILR is able to achieve this result by validating only 11% of the photos (i.e., 204 out of 1784).

TABLE 8.3
Simulation setup for the ILR scheme

Parameter	Value
Number of nodes	200
% of tasks with false location claims	10, 15, 30, 45, 60
Bluetooth transmission range	10m
Simulation time	2hrs
User walking speed	1m/sec
Node density	2, 3, 4, 5
Bluetooth scan rate	1/min

8.3.1.5   ILR Simulations

This section presents the evaluation of the ILR scheme using the NS-2 network simulator. The two main goals of the evaluation are: (1) Estimate the right percentage of photo tasks needed in Phase 1 to bootstrap the ILR scheme, and (2) Quantify the ability of ILR to detect false location claims at various node densities.

Simulation Setup:

The simulation setup parameters are presented in Table 8.3. Given a simulation area of 100m x 120m, the node degree (i.e., average number of neighbors per user) is slightly higher than 5. Talasila et al. varied the simulation area to achieve node degrees of 2, 3, and 4. The authors consider low walking speeds (i.e., 1m/sec) for collecting photos. In these simulations, the authors considered all tasks as photo tasks. A photo task is executed every minute by each node. Photo tasks are distributed evenly across all nodes. Photo tasks with false location claims are also distributed evenly across several malicious nodes. The authors assume the photo tasks in ILR's phase 1 are manually validated.

After executing the simulation scenarios described below, the authors collected each photo task's time, location, and Bluetooth scan. As per simulation settings, there will be 120 completed photo tasks per node at the end of the simulation (i.e., 24,000 total photo tasks for 200 nodes). Over this collected data, Talasila et al. applied the ILR validation scheme to detect false location claims.

Simulation Results:

Varying percentage of false location claims. In this set of experiments, the authors varied the percentage of photo tasks with false location claims. The resulting graph, plotted in Figure 8.6, has multiple curves as a function of the percentage of photo tasks submitting false locations. This graph is plotted to gain insights on what will be the right percentage of photo tasks needed in Phase 1 to bootstrap the ILR scheme. Next, we analyze Figure 8.6:

- **Low count of malicious tasks submitted:** When 10% of total photo tasks are submitting false location, Figure 8.6 shows that just by using 10% of the total photo tasks validated in Phase 1, the ILR scheme can detect 55% of the false location claims. This figure also shows that in order to detect more false claims, we can use up to 40% of the total photo tasks in Phase 1 to detect 80% of the false location tasks. Finally, Figure 8.6 shows that increasing the percentage of validated photo tasks above 40% will not help much as the percentage of detected false tasks remains the same.

- **High count of malicious tasks submitted:** When 60% of the total photo tasks are submitting false locations, Figure 8.6 shows that ILR can still detect 35% of the false claims by using 10% of the total photo tasks in Phase 1. But in this case, the ILR scheme requires more validated photo tasks (70%) to catch 75% of the false claims. This is because by increasing the number of malicious tasks, the co-location data is reduced and therefore ILR cannot extend trust to more location claims in its Phase 2.

Therefore, Talasila et al. concluded that the right percentage of photo tasks needed to bootstrap the ILR scheme is proportional to the expected false location claims (which can be predicted using the history of the users' participation).

Node density impact on the ILR scheme. In this set of experiments, the authors assume that 10% of the total photo tasks are submitting false locations. In Figure 8.7, Talasila et al. analyzed the impact of node density on the ILR scheme. The authors seek to estimate the minimum node density required to achieve highly connected graphs, to extend the location trust transitively to more co-located nodes.

- **High Density:** When simulations are run with node density of 5, Figure 8.7 shows the ILR scheme can detect the highest percentage (>85%) of the false location claims. The figure also shows similarly high results even for a node density of 4.

- **Low Density:** When simulations are run with node density of 2, we can see that the ILR scheme can still detect 65% of the false location tasks using 50% of the total photo tasks in Phase 1. For this node density, even after increasing the number of validated photo tasks in Phase 1, the percentage of detected false claims does not increase. This is because of there are fewer co-located users at low node densities.

Therefore, the authors concluded that the ILR scheme can efficiently detect false claims with a low number of manual validations, even for low node densities.

8.3.2 The LINK Scheme for Real-time Location Data Validation

This section describes LINK (Location authentication through Immediate Neighbors Knowledge) [154], a secure location authentication protocol in which users help authenticate each other's location claims. LINK associates trust scores with users, and nearby mobile devices belonging to users with high trust scores play similar roles with the trusted beacons/base stations in existing location-authentication solutions. The main idea is to leverage the neighborhood knowledge available through short-range wireless technologies, such as Bluetooth, which is available on most cell phones, to verify if a user is in the vicinity of other users with high trust scores.

Assumptions:

This section defines the interacting entities in the environment, and the assumptions the authors make about the system for LINK protocol. The interacting entities in the system are:

- *Claimer*: The mobile user who claims a certain location and subsequently has to prove the claim's authenticity.

- *Verifier*: A mobile user in the vicinity of the claimer (as defined by the transmission range of the wireless interface, which is Bluetooth in the implementation). This user receives a request from the claimer to certify the claimer's location and does so by sending a message to the LCA.

- *Location Certification Authority (LCA)*: A service provided in the Internet that can be contacted by location-based services to authenticate claimers' location. All mobile users who need to authenticate their location are registered with the LCA.

- *Location-Based Service (LBS)*: The service that receives the location information from mobile users and provides responses as a function of this location.

Talasila et al. assume that each mobile device has means to determine its location. This location is considered to be approximate, within typical GPS or other localization systems limits. The authors assume the LCA is trusted and the communication between mobile users and the LCA occurs over secure channels, e.g., the communication is secured using SSL/TLS. The authors also assume that each user has a pair of public/private keys and a digital certificate from a PKI. Similarly, the authors assume the LCA can retrieve and verify the certificate of any user. All communication happens over the Internet, except the short-range communication between claimers and verifiers.

The authors chose Bluetooth for short-range communication in LINK because of its pervasiveness in cell phones and its short transmission range (10m), which provides good accuracy for location verification. However, LINK can leverage WiFi during its initial deployment in order to increase the network density. This solution trades off location accuracy for number of verifiers.

LCA can be a bottleneck and a single point of failure in the system. To address these issues, standard distributed systems techniques can be used to improve the LCA's scalability and fault tolerance. For example, an individual LCA server/cluster can be assigned to handle a specific geographic region, thus reducing the communication overhead significantly (i.e., communication between LCA servers is only required to access a user's data when she travels away from the home region). LINK also needs significant memory and storage space to store historic data about each pair of users who interact in the system. To alleviate this issue, a distributed implementation of the LCA could use just the recent history (e.g., past month) to compute trust score trends, use efficient data-intensive parallel computing frameworks such as *Hadoop* [14] to pre-compute these trends offline, and employ distributed caching systems such as *Memcached* [20] to achieve lower latency for authentication decisions.

Adversarial Model:

Any claimer or verifier may be malicious. When acting individually, malicious claimers may lie about their location. Malicious verifiers may refuse to cooperate when asked to certify the location of a claimer and may also lie about their own location in order to slander a legitimate claimer. Additionally, malicious users may perform stronger attacks by colluding with each other in order to verify each other's false claims. Colluding users may also attempt two classic attacks: mafia fraud and terrrorist fraud [64].

Talasila et al. do not consider selfish attacks, in which users seek to reap the benefits of participating in the system without having to expend their own resources (e.g., battery). These attacks are solved by leveraging the centralized nature of LCA, which enforces a tit-for-tat mechanism, similar to those found in P2P protocols such as BitTorrent [57], to incentivize nodes to participate in verifications. Only users registered with the LCA can participate in the system as claimers and verifiers. The tit-for-tat mechanism requires the verifiers to submit verifications in order to be allowed to submit claims. New users are allowed to submit a few claims before being requested to perform verifications.

Finally, the authors rely on the fact that a user cannot easily obtain multiple user IDs because the user ID is derived from a user certificate, and obtaining digital certificates is not cheap; this deters Sybil attacks [67]. Further, techniques such as [127, 133], complimentary to the LINK protocol, can be used to address these attacks.

8.3.2.1   Protocol Design

This section presents the basic LINK operation, describes the strategies used by LCA to decide whether to accept or reject a claim, and then details how trust scores and verification history are used to detect strong attacks from malicious users who change their behavior over time or collude with each other.

Basic Protocol Operation:

All mobile users who want to use LINK must register with the LCA. During registration, the LCA generates a userID based on the user's digital certificate. Users of the system do not have to register with the LBS because they submit their LCA-generated userID to the LBS together with their requests. By not requiring the users to register with each LBS, the authors simplified the protocol.

At registration time, the LCA assigns an initial trust score for the user (which can be set to a default value or assigned based on other criteria). Trust scores are maintained and used by the LCA to decide the validity of location claims. A user's trust score is additively increased when her claim is successfully authenticated, and multiplicatively decreased otherwise in order to discourage malicious behavior. This policy of updating the scores is demonstrated to work well for the studied attacks, as shown in Section 8.3.2.4. The values of all trust score increments, decrements, and thresholds are presented in the same section. A similar trust score updating policy has been shown to be effective in P2P networks as well [56].

LCA also maintains a verification count of each user to determine whether the user is participating in verifications or simply using the system for her own claims. A user needs to perform at least VC_{th} verifications in order to be allowed to submit one claim (VC_{th} is the verification count threshold). Each time a verifier submits a verification, her verification count is incremented by 1, and each time a claimer submits a request to LCA, her verification count is decreased by VC_{th}. If a claimer's verification count reaches below VC_{th}, the claimer is informed to participate in other claimers' verifications and her claims are not processed until her verification count reaches above VC_{th}.

Figure 8.8 illustrates the basic LINK operation. The pseudo-code describing the actions of claimers, verifiers, LCA, and LBS is presented in Algorithm 2. LINK messages are signed. When the authors say that a protocol entity sends a signed message, it means that the entity computes a digital signature over the entire message and appends this signature at the end of the message. In step 1, a user (the claimer) wants to use the LBS and submits her location (Claimer Pseudo-code, line 5). The LBS then asks the claimer to authenticate her location (step 2) (LBS Pseudo-code, line 3). In response, the claimer will send a signed message to LCA (step 3) which consists of *(userID, serviceID, location, seq-no, serviceID, verifiers' IDs)* (Claimer Pseudo-code, line 12). The sequence number (*seq-no*) is used to protect against replay attacks (to be discussed in Section 8.3.2.3). The *serviceID* is an identifier of the LBS. The verifiers' IDs consists of the list of verifiers discovered by the claimer's Bluetooth scan; in this way, LCA will ignore the certification replies received from any other verifiers (the purpose of this step is to defend against mafia fraud attacks as detailed in Section 8.3.2.3). Furthermore, the LCA timestamps and stores each newly received claim.

The claimer then starts the verification process by broadcasting to its neighbors a location certification request over the short-range wireless interface (step 4). This message is signed and consists of *(userID, serviceID, location, seq-no)*, with the same sequence number as the claim in step 3. The neighbors who receive the message, acting as verifiers for the claimer, will send a signed certification reply message to LCA (step 5) (Verifier Pseudo-code, line 8). This message consists of *(userID, location, certification-request)*, where the userID and location are those of the verifier and certification-request is the certification-request broadcasted by the claimer. The certification-request is included to allow the LCA to match the claim and its certification messages. Additionally, it proves that indeed the certification-reply is in response to the claimer's request.

The LCA waits for the certification reply messages for a short period of time and then starts the decision process (described next in Section 8.3.2.2). Finally, the LCA informs the LBS about its decision (step 6) (LCA Pseudo-code, line 9), causing the LBS to provide or deny service to the claimer (LBS Pseudo-code, line 8).

8.3.2.2 LCA Decision Process

In the following, we present the pseudo-code (Algorithm 3) and description of the LCA decision process. For the sake of clarity, this description skips most of the details regarding the use of historical data when making decisions, which are presented in Section 8.3.2.2.

Claimer lies:

The LCA first checks the user's spatio-temporal correlation by comparing the currently claimed location with the location of the user's previously recorded claim (lines 1–3 in the algorithm). If it is not physically possible to move between these locations in the time period between the two claims, the new claim is rejected.

If the claimer's location satisfies the spatio-temporal correlation, the LCA selects only the "good" verifiers who responded to the certification request and who are in the list of verifiers reported by the claimer (lines 5–12). These verifiers must include in their certification reply the correct certification request signed by the claimer (not shown in the code) and must satisfy the spatio-temporal correlation themselves. Additionally, they must have trust scores above a certain threshold. The authors only use "good" verifiers because verifiers with low scores may be malicious and may try to slander the claimer. Nevertheless, the low score verifiers respond to certifica-

---

Algorithm 2 LINK Pseudo-Code

Claimer Pseudo-code:

1: // KC_{pr} = private key of claimer
2: // KC_{pu} = public key of claimer
3: // seq-no = provided by LCA before each claim
4: // LBS-request = [userID, location]
5: send(LBS, LBS-request)
6: receive(LBS, response)
7: **if** *response == Authenticate* **then**
8:    certification-request = [userID, serviceID, location, seq-no];
9:    certification-request += sign(KC_{pr}, certification-request);
10:   verifiers = BluetoothDiscoveryScan();
11:   signed-verifiers = verifiers + sign(KC_{pr}, verifiers);
12:   send(LCA, certification-request, signed-verifiers);
13:   broadcast(verifiers, certification-request);
14:   receive(LBS, response);

Verifier Pseudo-code:

1: // KV_{pr} = private key of verifier
2: // KV_{pu} = public key of verifier
3: // Verifier enables Bluetooth Radio to discoverable mode
4: claimer = BluetoothDiscovered();
5: receive(claimer, certification-request);
6: certification-reply = [userID, location, certification-request];
7: certification-reply += sign(KV_{pr}, certification-reply);
8: send(LCA, certification-reply);
9: // Potential request from LCA to authenticate its location

LCA Pseudo-code:

1: receive(claimer, certification-request, verifiers)
2: **if** $verify(KC_{pu}, certification-request)$ **and** $verify(KC_{pu}, verifiers)$ **then**
3:    **for** $v = 1$ to verifiers.size() **do**
4:       receive(v, certification-reply[v]);
5:       **if** $verify(KV_{pu}, certification - reply[v])$ **and** $verify(KC_{pu}, getCertificationRequest(certification - reply[v]))$ **then**
6:          storeDataForDecision(certification-request, certification-reply)
7: decision = decisionProcess(claimer, getClaimLocation(certification-request));
8: LBS = getServiceID(certification-request);
9: send(LBS, decision);

LBS Pseudo-code:

1: receive(claimer, LBS-request);
2: **if** *locationAuthentication == Required* **then**
3:    send(claimer, Authenticate);
4:    receive(LCA, decision);
5:    **if** *decision == Reject* **then**
6:       send(claimer, ServiceDenied)
7:       **return**
8: send(claimer, response);

---

tion requests in order to be allowed to submit their own certification claims (i.e., tit-for-tat mechanism) and, thus, potentially improve their trust scores.

Notation of Algorithm 3:

```
c: claimer
V = {v_0,v_1,...v_n}: Set of verifiers for claimer c
N_set: Set of verifiers who do not agree with c's location claim
v_i: The i-th verifier in V
T_{v_i}: Trust score of verifier v_i
T_c: Trust score of claimer
W_{v_i}: Weighted trust score of verifier v_i
L_c: Location claimed by claimer
L_{v_i}: Location claimed by verifier v_i
IND_{tr}: Individual threshold to eliminate the low-scored verifiers
AVG_{tr}: Average threshold to ensure enough difference in averages
VRF_{cnt}=0: Variable to hold recursive call count
INC=0.1: Additive increment
DEC=0.5: Multiplicative decrement
secVer[]: Array to hold the response of second level verifications
```

Colluding users help claimer:

After selecting the "good" verifiers, the LCA checks if they are colluding with the claimer to provide false verifications (lines 13–15), and it rejects the claim if that is the case. This *collusion check* is described in detail in Section 8.3.2.2 under the paragraph titled, "Colluding users verification."

Contradictory verifications:

If the LCA does not detect collusion between the claimer and verifiers, it accepts or rejects the claim based on the difference between the sums of the trust scores of the two sets of verifiers (lines 16–23), those who agree with the location submitted by the claimer (Y_{sum}), and those who do not (N_{sum}). Of course, the decision is easy as long as all the verifiers agree with each other. The difficulty comes when the verifiers do not agree with each other. This could be due to two causes: malicious individual verifiers, or verifiers colluding with the claimer, who have escaped detection.

If the difference between the trust score sums of two sets of verifiers is above a certain threshold, the LCA decides according to the "winning" set. If it is low, the LCA does not make a decision yet. It continues by checking the trust score trend of the claimer (lines 24–27): if this trend is poor, with a pattern of frequent score increases and decreases, the claimer is deemed malicious and the request rejected. Otherwise, the LCA checks the score trends of the verifiers who disagree with the claimer (lines 28). If these verifiers are deemed malicious, the claim is accepted. Otherwise, the claim is ignored, which forces the claimer to try another authentication later.

Algorithm 3 Decision Process Pseudo-Code

---

decisionProcess(c, $\mathbf{L}_c$):
:run to validate the location $\mathbf{L}_c$ claimed by c

1: **if** $SpatioTempCorrelation(c) == FALSE$ **then**
2:     $T_c = T_c * DEC$
3:     **return** Reject
4: VRF_{cnt}++
5: **if** $hasNeighbors(c) == TRUE$ **then**
6:     **for** $i = 0$ to n **do**
7:         **if** $SpatioTempCorrelation(v_i) == FALSE$ **then**
8:             $T_{v_i} = T_{v_i} * DEC$
9:             remove v_i from set V
10:         $W_{v_i} = $ getUpdatedWeightScore(v_i,c)
11:         **if** $W_{v_i} \leq IND_{tr}$ **then**
12:             remove v_i from set V
13:     **if** $V.notEmpty()$ **then**
14:         **if** $checkCollusion(V,c) == TRUE$ **then**
15:             **return** Reject
16:         (absAVGDiff,Y_{sum},N_{sum}) = getTrustScore(V,c)
17:         **if** $absAVGDiff \geq AVG_{tr}$ **then**
18:             **if** $Y_{sum} \geq N_{sum}$ **then**
19:                 $T_c = T_c + INC$
20:                 **return** Accept
21:             **else**
22:                 $T_c = T_c * DEC$
23:                 **return** Reject
24:         **else**
25:             **if** $trend(c) == POOR$ **then**
26:                 $T_c = T_c * DEC$
27:                 **return** Reject
28:             **else if** $verifyScoreTrends(N_{set}) == POOR$ **then**
29:                 **if** $T_c \leq IND_{tr}$ **then**
30:                     **return** Ignore
31:                 **else**
32:                     $T_c = T_c - INC$
33:                     **return** Accept
34:             **else if** $VRF_{cnt} == 2$ **then**
35:                 **return** Ignore {//2nd level claim is ignored}
36:             **else**
37:                 **for** $i = 0$ to $N_{set}.size()$ **do**
38:                     secVer[i] = decisionProcess(v_i,L_{v_i})
39:                 **if** $Majority(secVer) == Ignore$ or $Reject$ **then**
40:                     $T_c = T_c + INC$
41:                     **return** Accept
42:                 **else**
43:                     $T_c = T_c * DEC$
44:                     **return** Reject
45: **if** $VRF_{cnt} == 2$ **then**
46:     **return** Ignore {//2nd level claim is ignored}
47: **else if** $trend(c) == POOR$ **then**
48:     $T_c = T_c * DEC$
49:     **return** Reject
50: **else if** $T_c \leq IND_{tr}$ **then**
51:     **return** Ignore
52: **else**
53:     $T_c = T_c - INC$
54:     **return** Accept

Note that even if the claim is accepted in this phase, the trust score of the claimer is preventively decremented by a small value (lines 32–33). In this way, a claimer who submits several claims that are barely accepted will receive a low trust score over time; this trust score will prevent future "accepts" in this phase (lines 29-30) until her trust scores improves.

If the trend scores of both the claimer and the verifiers are good, the verifiers are challenged to authenticate their location (lines 34–44). This second level verification is done through a recursive call to the same *decisionProcess()* function. This function is invoked for all verifiers who do not agree with the claimer (lines 37–38). If the majority of these verifiers cannot authenticate their location (i.e., Ignore or Reject answers), the claim is accepted (lines 39–41). Otherwise, the claim is rejected. The VRF_{cnt} variable is used to keep track of the recursive call count. Since only one additional verification level is performed, the function returns when its value is 2.

Claimer with no verifiers:

The LCA deals with the case when no "good" verifiers are found to certify the claim in lines 45–54 of the algorithm (this includes no verifiers at all). If the claimer's trust score trend is good and her trust score is higher than a certain threshold, the claim is accepted. In this situation, the claimer's trust score is decreased by a small value $INC = 0.1$, as shown at line 53 in Algorithm 3, to protect against malicious claimers who do not broadcast a certification request to their neighbors when they make a claim. Over time, a user must submit claims that are verified by other users; otherwise, all her claims will be rejected.

Use of Historical Data in LCA Decision

The LCA maintains for each user the following historical data: (1) all values of the user's trust score collected over time, and (2) a list of all users who provided verifications for this user together with a verification count for each of them. These data are used to detect and prevent attacks from malicious users who change their behavior over time or who collude with each other.

Trust score trend verification. The goal of this verification is to analyze the historical trust values for a user and find malicious patterns. This happens typically when there are no good verifiers around a claimer or when the verifiers contradict each other, with no clear majority saying to accept or reject the claim.

For example, a malicious user can submit a number of truthful claims to improve her trust score and then submit a malicious claim without broadcasting a certification request to her neighbors. Practically, the user claims to have no neighbors. This type of attack is impossible to detect without verifying the historical trust scores. To prevent such an attack, the LCA counts how many times a user's trust score has been decreased over time. If this number is larger than a certain percentage of the total number of claims issued by that user (10% in the implementation), the trend is considered malicious. More complex functions or learning methods could be used, but this simple function works well for many types of attacks, as demonstrated by the experiments.

Colluding users verification. Groups of users may use out-of-band communication to coordinate attacks. For example, they can send location certification messages to LCA on behalf of each other with agreed-upon locations. To mitigate such attacks, the LCA maintains an NxN matrix M that tracks users certifying each other's claims (N is the total number of users in the system). $M[i][c]$ counts how many times user i has acted as verifier for user c. The basic idea is that colluding users will frequently certify each other's claims compared with the rest of the users in the system. However, identifying colluding users based solely on this criterion will not work, because a spouse or a colleague at the office can very frequently certify the location of certain users. Furthermore, a set of colluding malicious users can use various permutations of subsets of malicious verifiers to reduce the chances of being detected.

Therefore, Talasila et al. propose two enhancements. First, the LCA algorithm uses weighted trust scores for verifiers with at least two verifications for a claimer. The weighted trust score of a verifier v is $W_v = T_v / \log_2(M[i][c])$, where T_v is the actual trust score of v. The more a user certifies another user's claims, the less its certifying information will contribute in the LCA decision. Talasila et al. chose a log function to induce a slower decrease of the trust score as the count increases. Nevertheless, a small group of colluding users can quickly end up with all their weighted scores falling below the threshold for "good" users, thus stopping the attack.

This enhancement is used until enough verification data is collected. Then, it is used in conjunction with the second enhancement, which discriminates between colluding malicious users and legitimate users who just happen to verify often for a claimer. LINK rejects a claim if the following conditions are satisfied for the claimer:

1. The number of claims verified by each potentially colluding user is greater than a significant fraction of the total number of claims issued by the claimer, and

2. The number of potentially colluding users who satisfy the first condition is greater than a significant fraction of the total number of verifiers for the claimer.

The pseudo-code for this function is presented in Algorithm 4. The function uses a dynamic threshold, W_{max}, which is a percentage of the total number of claims issued by the claimer over time (in the implementation, this percentage, β, is set to 30%). Since W_{max} is dynamically set, the algorithm can adapt its behavior over time. The function then computes the percentage of the verifiers who already reached W_{max} (lines 3–6). If a significant number of verifiers reached this threshold, the algorithm considers them malicious and punishes them together with the claimer. Simulation results presented in Section 8.3.2.4

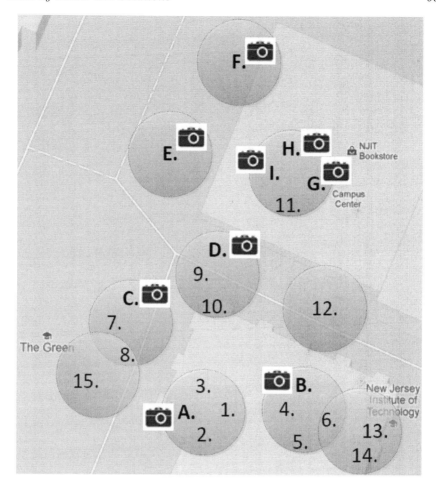

FIGURE 8.4
Example of McSense collected photo tasks [A–I] and sensing tasks [1–15] on
the campus map, grouped using Bluetooth discovery co-location data.

demonstrate the advantage of punishing the verifiers as well vs. a method that
would punish only the claimer.

A higher percentage of users verifying often for the same claimer is a
strong indication of malicious behavior (the parameter α, set to 10% in the
implementation, is used for this purpose). The underlying assumption is that
a legitimate user going about her business is verified by many users over time,
and only a few of them would verify often (e.g., family, lab mates).

Lines 7–13 show how the decision is made. If the number of potentially
colluding verifiers is greater than α, the claimer and those verifiers are punished. Note that the authors do not punish a verifier who did not participate
in verifications for this claimer since the last time she was punished (line 10).

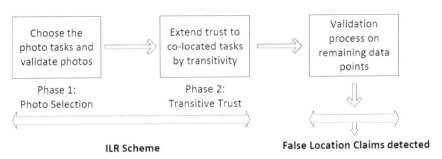

FIGURE 8.5
The phases of the ILR scheme.

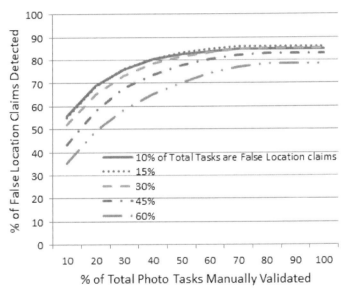

FIGURE 8.6
ILR performance as a function of the percentage of photos manually validated in Phase 1. Each curve represents a different percentage of photos with fake locations.

In this way, the verifiers can redeem themselves, but at the same time, their contribution is still remembered in M. Finally, as shown in lines 14–18, if the percentage of potentially colluding users is less than α, the counts for those users are reset to allow them to have a greater contribution in future verifications for the claimer (this is correlated with the weighted trust score described previously).

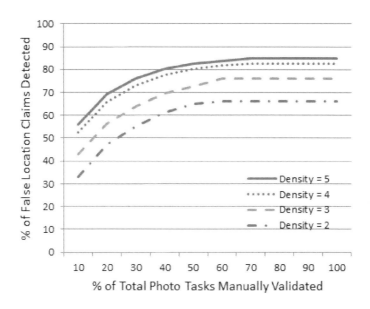

FIGURE 8.7

ILR performance as a function of the percentage of photos manually validated in Phase 1. Each curve represents a different network density, shown as the average number of neighbors per node.

8.3.2.3 Security Analysis

LINK's goal is to prevent malicious users from claiming an incorrect location and to accept truthful location claims from legitimate users. The decision made by the LCA to accept or reject a claim relies on the trust scores of the users involved in this claim (i.e., claimer and verifiers). Thus, from a security perspective, the protocol's goal is to ensure that *over time* the trust score of malicious users will decrease, whereas the score of legitimate users will increase. LINK uses an additive increase and multiplicative decrease scheme to manage trust scores in order to discourage malicious behavior.

There are certain limits to the amount of adversarial presence that LINK can tolerate. For example, LINK cannot deal with an arbitrarily large number of malicious colluding verifiers supporting a malicious claimer because it becomes very difficult to identify the set of colluding users. Similarly, LINK cannot protect against users who accumulate high scores and very rarely issue false claims while pretending to have no neighbors (i.e., the user does not broadcast a certification request). An example of such a situation is a "hit and run" attack, when the user does not return to the system after issuing a false claim. This type of behavior cannot be prevented even in other real-world systems that rely on user reputation, such as Ebay. Thus, Talasila et al. do

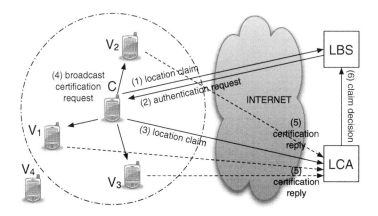

FIGURE 8.8
Basic protocol operation (where C = claimer, V_i = verifiers, LBS = location-based service, LCA = location certification authority).

not focus on preventing such attacks. Instead, the authors focus on preventing users that *systematically* exhibit malicious behavior. Up to a certain amount of adversarial presence, the simulation results in Section 8.3.2.4 show that the protocol is able to decrease, over time, the scores of users that exhibit malicious behavior consistently, and to increase the scores of legitimate users.

All certification requests and replies are digitally signed, thus the attacker cannot forge them, nor can she deny messages signed under her private key. Attackers may attempt simple attacks such as causing the LCA to use the wrong certification replies to verify a location claim. LINK prevents this attack by requiring verifiers to embed the certification request in the certification reply sent to the LCA. This also prevents attackers from arbitrarily creating certification replies that do not correspond to any certification request, as they will be discarded by the LCA.

Another class of attacks claims a location too far from the previously claimed location. In LINK, the LCA prevents these attacks by detecting it is not feasible to travel such a large distance in the amount of time between the claims.

The LCA's decision-making process is facilitated when there is a clear difference between the trust scores of legitimate and malicious users. This corresponds to a stage in which the user scores have stabilized (i.e., malicious scores have low scores and legitimate users have high scores). However, there may be cases when this score difference is not significant and it becomes challenging to differentiate between a legitimate verifier vouching against a malicious claimer and a malicious verifier slandering a legitimate claimer. In this case, the LCA's decision relies on several heuristic rules. The true nature of a user (malicious or legitimate) may be reflected in the user's score trend

---

Algorithm 4 Collusion Check Pseudo-Code

Notation:
M: NxN matrix for keeping count of certifications for pairs (claimer, verifier)
W_{max}: Threshold for count w
k: Number of verifiers with $w \geq W_{max}$ for claimer c
V: Set of active verifiers for claimer c
T_c: Trust score of claimer
T_{v_i}: Trust score of i-th verifier in V
m: Number of verifiers with $w > 0$ for claimer c
$NumCl_c$: Total number of claims made by claimer c
W_{rst}: reset value

checkCollusion(V, c):
```
 1: Wmax = β * NumClc
 2: for i = 0 to M.size do
 3:     if M[i][c] ≥ Wmax then
 4:         k++;
 5:     if M[i][c] > 0 then
 6:         m++;
 7: if k/m ≥ α then
 8:     Tc = Tc * DEC
 9:     for i = 0 to M.size do
10:         if (M[i][c] ≥ Wmax) and ((i ∈ V) or (i.punished == FALSE)) then
11:             Ti = Ti * DEC
12:             i.punished = TRUE
13:     return TRUE
14: else
15:     for i = 0 to V.size do
16:         if M[V[i]][c] ≥ Wmax then
17:             M[V[i]][c] = Wrst
18:     return FALSE
```

---

and the LCA can decide based on the score trends of the claimer and verifiers. The LCA may also potentially require the verifiers to prove their location. This additional verification can reveal malicious verifiers that are certifying a position claim (even though they are not in the vicinity of the claimed position), because the verifiers will not be able to prove their claimed location.

Replay Attack. Attackers may try to slander other honest nodes by intercepting their certification requests and then replaying them at a later time in a different location. However, the LCA is able to detect that it has already processed a certification request (extracted from a certification reply) because each such request contains a sequence number and the LCA maintains a record of the latest sequence number for each user. Thus, such duplicate requests will be ignored.

Individual Malicious Claimer or Verifier Attacks. We now consider individual malicious claimers that claim a false location. If the claimer follows the protocol and broadcasts the certification request, the LCA will reject the claim because the claimer's neighbors provide the correct location and prevail over the claimer. However, the claimer may choose not to broadcast the certification request and only contact the LCA. If the attacker has a good trust score, she will get away with a few false claims. The impact of this attack is limited because the attacker trust score is decreased by a small decrement for each such claim, and she will soon end up with a low trust score; consequently, all future claims without verifiers will be rejected. Accepting a few false claims is a trade-off the authors adopted in LINK in order to accept location claims from legitimate users that occasionally may have no neighbors.

An individual malicious verifier may slander a legitimate user who claims a correct location. However, in general, the legitimate user has a higher trust score than the malicious user. Moreover, the other (if any) neighbors of the legitimate user will support the claim. The LCA will thus accept the claim.

Colluding Attack. A group of colluding attackers may try to verify each other's false locations using out-of-band channels to coordinate with each other. For example, one attacker claims a false position and the other attackers in the group support the claim. LINK deals with this attack by recording the history of verifiers for each claimer and gradually decreasing the contribution of verifiers that repeatedly certify for the same claimer (see Section 8.3.2.2). Even if this attack may be successful initially, repeated certifications from the same group of colluding verifiers will eventually be ignored (as shown by the simulations in Section 8.3.2.4).

Mafia Fraud. In this attack, colluding users try to slander honest claimers without being detected, which may lead to denial-of-service. For example, a malicious node M_1 overhears the legitimate claimer's certification request and relays it to a remote collaborator M_2; M_2 then re-broadcasts this certification request pretending to be the legitimate claimer. This results in conflicting certification replies from honest neighbors of the legitimate claimer and honest neighbors of M_2 from a different location. This attack is prevented in LINK because the LCA uses the list of verifiers reported by the legitimate claimer from its Bluetooth scan. Therefore, LCA ignores the certification replies of the extra verifiers who are not listed by the legitimate claimer. These extra verifiers are not punished by LCA, as they are being exploited by the colluding malicious users. Furthermore, it is difficult for colluding users to follow certain users in order to succeed in such an attack.

Limitations and Future Work. The thresholds in the protocol are set based on the expectations of normal user behavior. However, they can be modified or even adapted dynamically in the future.

LINK was designed under the assumption that users are not alone very often when sending the location authentication requests. As such, it can lead to significant false positive rates for this type of scenario. Thus, LINK is best applicable to environments in which user density is relatively high.

Terrorist fraud is another type of attack in which one attacker relays the certification request to a colluding attacker at a different location, in order to falsely claim the presence at that different location. For example, a malicious node M_1 located at location L_1 relays its certification request for location L_2 to collaborator M_2 located at L_2. M_2 then broadcasts M_1's request to nearby verifiers. Verifiers certify this location request, and as a result the LCA falsely believes that M1 is located at L_2. This attack is less useful in practice and is hard to mount, as it requires one of the malicious users to be located at the desired location. From a philosophical perspective, attacker M_1 *has a physical presence* at the falsely claimed location through its collaborator M_2, so this attack is arguably not a severe violation of location authentication. The attack could be prevented by using distance bounding protocols [150, 47, 138]. However, applying such protocols is non-trivial. Two major issues are: (a) These protocols consider the verifiers as trusted whereas LINK verifiers are not trusted, and (b) special hardware (i.e., radio) not available on current phones may be needed to determine distances with high accuracy.

In addition, a group of colluding attackers may attempt a more sophisticated version of the terrorist fraud attack, in which the malicious collaborators share their cryptographic credentials to sign the false location claims for each other. In practice, this scenario is unlikely because a malicious user will be reluctant to share her credentials with another malicious user [138].

The authors implicitly assume that all mobile devices have the same nominal wireless transmission range. One can imagine ways to break this assumption, such as using non-standard wireless interfaces that can listen or transmit at higher distances, such as the BlueSniper rifle from DEFCON '04. In this way, a claimer may be able to convince verifiers that she is indeed nearby, while being significantly farther away. Such attacks can also be prevented similar to the above solution for terrorist fraud using distance bounding protocols [150, 47, 138].

Location privacy could be an issue for verifiers. Potential solutions may include rate limitations (e.g., number of verifications per hour or day), place limitations (e.g., do not participate in verifications in certain places), or even turning LINK off when not needed for claims. However, the tit-for-tat mechanism requires the verifiers to submit verifications in order to be allowed to submit claims. To protect verifier privacy against other mobile users in proximity, the verification messages could be encrypted as well.

8.3.2.4 Simulations

This section presents the evaluation of LINK using the ns-2 simulator. The two main goals of the evaluation are: (1) Measuring the false negative rate (i.e., percentage of accepted malicious claims) and false positive rate (i.e., percentage of denied truthful claims) under various scenarios, and (2) verifying whether LINK's performance improves over time as expected.

Simulation Setup:

The simulation setup parameters are presented in Table 8.4. The average number of neighbors per user considering these parameters is slightly higher than 5. Since the authors are interested in measuring LINK's security performance, not its network overhead, they made the following simplifying changes in the simulations. Bluetooth is emulated by WiFi with a transmission range of 10m. This results in faster transmissions as it does not account for Bluetooth discovery and connection establishment. However, the impact on security of

TABLE 8.4

Simulation setup for the LINK protocol

Parameter	Value
Simulation area	100m x 120m
Number of nodes	200
% of malicious users	1, 2, 5, 10, 15
Colluding user group size	4, 6, 8, 10, 12
Bluetooth transmission range	10m
Simulation time	210min
Node speed	2m/sec
Claim generation rate (uniform)	1/min, 1/2min, 1/4min, 1/8min
Trust score range	0.0 to 1.0
Initial user trust score	0.5
''Good'' user trust score threshold	0.3
Low trust score difference threshold	0.2
Trust score increment	0.1
Trust score decrement - common case	0.5
Trust score decrement - no neighbors	0.1

this simplification is limited due to the low walking speeds considered in these experiments. Section 8.3.2.6 will present experimental results on smartphones that quantify the effect of Bluetooth discovery and Piconet formation. The second simplification is that the communication between the LCA and the users does not have any delay; the same applies for the out-of band communication between colluding users. Finally, a few packets can be lost due to wireless contention because the authors did not employ reliable communication in their simulation. However, given the low claim rate, the impact of these packets is minimal.

To simulate users' mobility, Talasila et al. used the Time-variant Community Mobility Model (TVCM model) [90], which has the realistic mobility characteristics observed from wireless LAN user traces. Specifically, TVCM selects frequently visited communities (areas that a node visits frequently) and different time periods in which the node periodically re-appears at the same location. Talasila et al. use the following values of the TVCM model in the simulations: 5 communities, 3 periods, and randomly placed communities represented as squares having the edge length 20m. The TVCM features help in providing a close approximation of real-life mobility patterns compared to the often-used random waypoint mobility model (RWP). Nevertheless, to collect additional results, the authors ran simulations using both TVCM and RWP. For most experiments, they have seen similar results between the TVCM model and the RWP model. Therefore, they omit the RWP results. There is one case, however, in which the results for TVCM are worse than the results for RWP: It is the "always malicious individual verifiers," and this difference will be pointed out when we discuss this case.

Simulation Results:

Always malicious individual claimers. In this set of experiments, a certain number of non-colluding malicious users send only malicious claims; however, they verify correctly for other claims.

If malicious claimers broadcast certification requests, the false negative rate is always 0. These claimers are punished and, because of low trust scores, they will not participate in future verifications. For higher numbers of malicious claimers, the observed false positive rate is very low (under 0.1%), but not 0. The reason is that a small number of good users remain without neighbors for several claims and, consequently, their trust score is decreased; similarly, their trust score trend may seem malicious. Thus, their truthful claims are rejected if they have no neighbors. The users can overcome this rare issue if they are made aware that the protocol works best when they have neighbors.

If malicious claimers do not broadcast certification requests, a few of their claims are accepted initially because it appears that they have no neighbors. If a claimer continues to send this type of claim, her trust score falls below the "good" user threshold and all her future claims without verifiers are rejected. Thus, the false negative rate will become almost 0 over time. The false positive rate remains very low in this case.

Sometimes malicious individual claimers. In this set of experiments, a malicious user attempts to "game" the system by sending not only malicious claims but also truthful claims to improve her trust score. Talasila et al. have evaluated two scenarios: (1) Malicious users sending one truthful claim, followed by one false claim throughout the simulation, and (2) Malicious users sending one false claim for every four truthful claims. For the first 10 minutes of the simulation, they send only truthful claims to increase their trust score. Furthermore, these users do not broadcast certification requests to avoid being proved wrong by others.

Figure 8.9 shows that LINK quickly detects these malicious users. Initially, the false claims are accepted because the users claim to have no neighbors and have good trust scores. After a few such claims are accepted, LINK detects the attacks based on the analysis of the trust score trends and punishes the attackers.

Figure 8.10 illustrates how the average trust score of the malicious users varies over time. For the first type of malicious users, the multiplicative decrease followed by an additive increase cannot bring the score above the "good" user threshold; hence, their claims are rejected even without the trust score trend analysis. However, for the second type of malicious users, the average trust score is typically greater than the "good" user threshold. Nev-

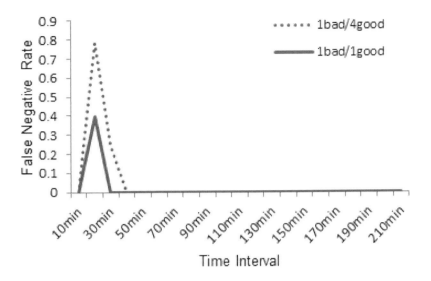

FIGURE 8.9
False negative rate over time for individual malicious claimers with mixed behavior. The claim generation rate is 1 per minute, 15% of the users are malicious, and average speed is 1m/s.

ertheless, they are detected based on the trust score trend analysis. In these simulations, the trust score range is between 0 and 1, i.e., additive increase of a trust score is done until it reaches 1, then it is not incremented anymore. It stays at 1, until there is a claim rejection or colluding verifier punishment.

Always malicious individual verifiers. The goal of this set of experiments is to evaluate LINK's performance when individual malicious verifiers try to slander good claimers. In these experiments, there are only good claimers, but a certain percentage of users will always provide malicious verifications.

Figure 8.11 shows that LINK is best suited for city environments where user density of at least 5 or higher can be easily found. The authors observed that for user density less than 4 LINK cannot afford more than 10% malicious verifiers, and for user density of 3 or less LINK will see high false positive rates due to no verifiers or due to the malicious verifiers around the claimer.

From Figure 8.12, we observe that LINK performs well even for a relatively high number of malicious verifiers, with a false positive rate of at most 2%. The 2% rate happens when a claimer has just one or two neighbors and those neighbors are malicious. However, a claimer can easily address this attack by re-sending a claim from a more populated area to increase the number of verifiers.

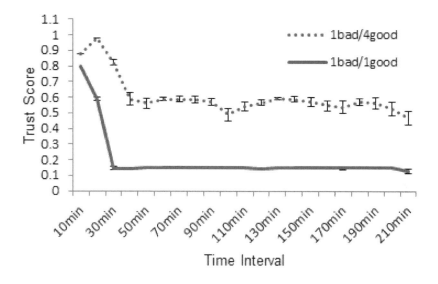

FIGURE 8.10

Trust score of malicious users with mixed behavior over time. The claim generation rate is 1 per minute, 15% of the users are malicious, and average speed is 1m/s. Error bars for 95% confidence intervals are plotted.

Of course, as the number of malicious verifiers increases, LINK can be defeated. Figure 8.13 shows that once the percentage of malicious users goes above 20%, the false positive rate increases dramatically. This is because the trust score of the slandered users decreases below the threshold and they cannot participate in verifications, which compounds the effect of slandering. This is the only result for which the authors observed significant differences between the TVCM model and the RWP model, with RWP leading to better results. This difference is due to the fact that nodes in same community in TVCM move together more frequently, which compounds the effect of slandering by malicious verifiers.

Colluding malicious claimers. This set of experiments evaluates the strongest attack the authors considered against LINK. Groups of malicious users collude, using out-of-band communication, to verify for each other. Furthermore, colluding users can form arbitrary verification subgroups; in this way, their collusion is more difficult to detect. To achieve high trust scores for the colluding users, the authors consider that they submit truthful claims for the first 30 minutes of the simulation. Then, they submit only malicious claims. As these are colluding users, there are no honest verifiers involved in these simulations.

Figure 8.14 shows that LINK's dynamic mechanism for collusion detection

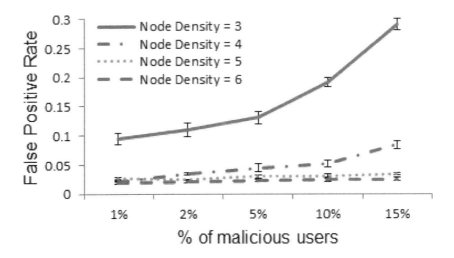

FIGURE 8.11
False positive rate as a function of the percentage of malicious verifiers for different node densities. The claim generation rate is 1 per minute and average speed is 1m/s. Error bars for 95% confidence intervals are plotted.

works well for these group sizes (up to 6% of the total nodes collude with each other). After a short period of high false negative rates, the rates decrease sharply and subsequently no false claims are accepted.

In LINK, all colluding users are punished when they are found to be malicious (i.e., the claimer and the verifiers). This decision could result in a few "good" verifiers being punished once in a while (e.g., family members). Figures 8.15 and 8.16 shows the false negative and positive rates, respectively, when punishing and not punishing the verifiers (i.e., the claimers are always punished). The authors observed that LINK takes a little more time to catch the colluding users while not punishing verifiers; at the same time, a small increase in the false positive rate is observed while punishing the verifiers. Since this increase in the false positive rate is not significant, the authors prefer to punish the verifiers in order to detect malicious claims sooner.

8.3.2.5   Implementation

The LINK prototype has been implemented and tested on Motorola Droid 2 smart phones installed with Android OS 2.2. These phones have 512 MB RAM, 1 GHz processor, Bluetooth 2.1, WiFi 802.11 b/g/n, 8 GB on board storage, and 8 GB microSD storage. Since the authors did not have data plans on their phones, all experiments were performed by connecting to the Internet over WiFi.

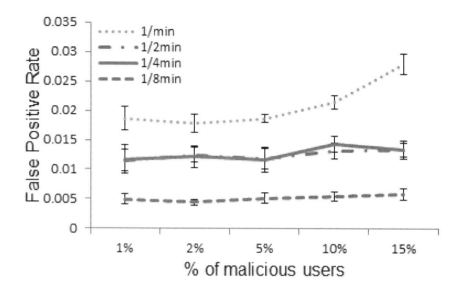

FIGURE 8.12

False positive rate as a function of the percentage of malicious verifiers for different claim generation rates. The average speed is 1m/s. Error bars for 95% confidence intervals are plotted.

The implementation consists of two main components: (1) an Android client side package that provides applications with a simple API to perform location authentication and allows users to start a background LINK verification process on the phones; (2) the LCA server implemented in Java. The communication between applications and LBSs can be done through any standard or custom API/protocol. To test LINK, Talasila et al. implemented, in Java, a Coupon LBS and its associated application that runs on smartphones (illustrated in Figure 8.17).

Client API:

We present the client API in the context of the Coupon LBS and its corresponding application. This service distributes location-based electronic discount coupons to people passing by a shopping mall. To prevent users located farther away from receiving these coupons, the service has to authenticate their location.

The corresponding application is implemented as an Android Application Project. The user is provided with a simple "Request Coupon" button as shown in Figure 8.17. The application submits the current location of the user to LBS and waits for an answer. Upon receiving the location authentication request from the LBS, the application invokes the *submit_claim* LINK API. An optimization done in the LINK package implementation was to limit the

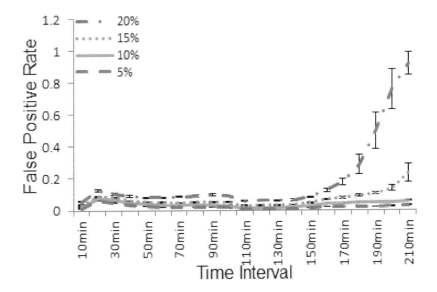

FIGURE 8.13
False positive rate over time for different percentages of malicious verifiers. The claim generation rate is 1 per minute and the average speed is 1m/s. Error bars for 95% confidence intervals are plotted.

Bluetooth discovery to 5.12s instead of the standard 10.24s. This improves the response time significantly and saves energy on the phones (as shown in Section 8.3.2.6) at the expense of rarely missing a neighboring phone. In [131], the authors show a single inquirer can locate 99% of all scanning devices within transmission range in 5.12s.

The Bluetooth discovery's callback function returns the list of discovered Bluetooth devices. This list may contain devices that do not belong to phones running LINK. Trying to establish Bluetooth connections with these devices will lead to useless consumption of resources and extra delays. Therefore, when *submit_claim* function contacts the LCA, it submits not only the location to be authenticated, but also the list of discovered Bluetooth devices. LCA answers with the assigned transaction ID for this claim and also provides the list of registered LINK-enabled devices. LINK on the phone will now establish a Bluetooth connection with each of these devices, sequentially, to send the claim certification request for the given transaction ID. Finally, the *submit_claim* function waits for the LCA answer, which is then returned to the application. If the response is positive, the application invokes another function to receive the coupon from the LBS; the LBS is informed of the authentication decision by the LCA directly.

For the verifier's side, which is not invoked from applications, the user has to start a LINK server that makes the Bluetooth listener active (i.e.,

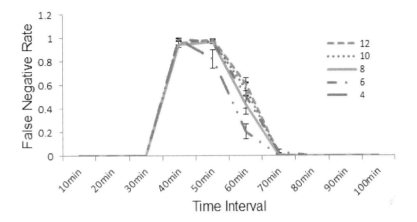

FIGURE 8.14
False negative rate over time for colluding users. Each curve is for a different colluding group size. Only 50% of the colluding users participate in each verification, thus maximizing their chances to remain undetected. The claim generation rate is 1 per minute and the average speed is 1m/s. Error bars for 95% confidence intervals are plotted.

puts Bluetooth in discoverable mode). This mode allows any claimer to find the phone in order to request the certification reply for their location claim. When a verifier's Bluetooth listener receives the certification request from a claimer, it invokes the *submit_verification* API. This function reads the claimer's message, generates the verification message, and sends it to the LCA. All the messages are signed using "SHA1withRSA" algorithm from the "java.security.Signature" package (1024 bits key size).

LCA Server:
    LCA is a multi-threaded server that maintains the claim transaction's hashmap, list of all user's details (ID, Bluetooth device address, RSA public key, trust score, etc.), and all users' weight matrices used in the decision process. One of the important implementation decisions is how long should a thread that received a claim wait for verifications to arrive[1]. This is necessary because some verifier phones may be turned off during the verification process, go out of the Bluetooth transmission range before the connection with the claimer is made, or even act maliciously and refuse to answer. This last example could lead to a denial of service attack on the LCA. Thus, the LCA cannot wait (potentially forever) until all expected verification messages

---

[1]The LCA knows the number of verifiers from the submitted claim message.

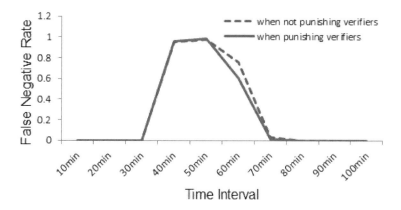

FIGURE 8.15
False negative rate over time when punishing and not punishing colluding verifiers. The size of the colluding group is 12, and 50% of these users participate in each verification. The claim generation rate is 1 per minute and the average speed is 1m/s.

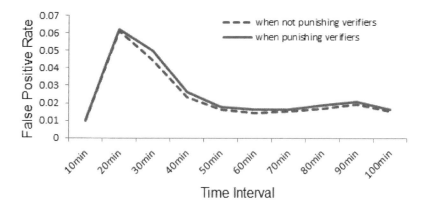

FIGURE 8.16
False positive rate over time when punishing and not punishing colluding verifiers. All parameters are the same as in Figure 8.15.

arrive. It needs a timeout, after which it makes the decision based on the verification received up to that moment.

Talasila et al. considered a waiting function linear in the number of verifiers. The linear increase is due to the sequential establishment of Bluetooth connections between the claimer and verifiers (i.e., they cannot be done in parallel). Since such a connection takes about 1.2s, the authors defined the waiting time $w = number\ of\ verifiers\ *\ 2s$, where 2s is an upper bound for the

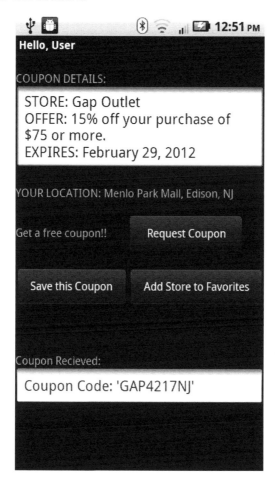

FIGURE 8.17
Coupon application on Android phone.

connection latency. However, this fixed waiting time could lead to long delays in situations when there are many verifiers and one or two do not answer at all. Therefore, the authors decided to adapt (i.e., reduce) this waiting time as a function of the number of received verifications. Upon each received verification, $w = w * 4/5$. The value of the reduction factor can be further tuned, but so far it worked well in the experiments.

Once all the verification replies are received or the timeout expires, the LCA processes the claim through the decision process algorithm. Finally, the LCA informs the claimer and the LBS about its decision.

TABLE 8.5

Latency table for individual LINK tasks

Task	Total time taken (s)
WiFi communication RTT	0.350
Bluetooth discovery	5.000
Bluetooth connection	1.200
Signing message	0.020
Verifying message	0.006

8.3.2.6   Experimental Evaluation

The main goals of these experiments are to understand the performance of LINK in terms of end-to-end response latency and power consumption when varying the number of claimers and verifiers. Additionally, Talasila et al. ran micro-benchmarks to understand the cost of individual tasks in LINK.

The authors used six Motorola Droid 2 smartphones for running the experiments [158]. The LCA and LBS server programs are deployed on Windows-based DELL Latitude D820 laptops, having Intel Core 2 CPU at 2.16GHz and 3.25GB RAM. Before starting the Coupon application, the smartphones' WiFi interfaces are switched on. Additionally, the background LINK processes that listen for incoming certification requests are started on each phone.

Measurements Methodology:

For latency, the roundtrip time of the entire LINK protocol (LINK RTT) is measured in seconds at the claimer mobile's coupon application program. For battery consumption, Power Tutor [186], available in the Android market, is used to collect power readings every second. The log files generated by PowerTutor are parsed to extract the CPU and WiFi power usage for the application's process ID. Separate tests are performed to benchmark the Bluetooth tasks as PowerTutor does not provide the Bluetooth radio power usage in its logs. All values are measured by taking the average for 50 claims for each test case.

Micro-Benchmark Results:

In these experiments, the authors used just two phones, one claimer and one verifier. Table 8.5 shows the latency breakdown for each individual task in LINK. Bluetooth discovery and Bluetooth connection are the tasks that took the major part of the response time. Note that the authors limited Bluetooth discovery to 5.12s, as explained in Section 8.3.2.5, to reduce the latency. From these results, they estimated that LINK latency is around 7s for one verifier; the latency increases linearly with the number of verifiers because the Bluetooth connections are established sequentially.

Table 8.6 shows the energy consumption breakdown for each task in LINK. The results show Bluetooth discovery consumes the most energy, while the

TABLE 8.6

Energy consumption for individual LINK tasks

Task	Energy Consumed (Joules)
WiFi communication RTT	0.100
Bluetooth discovery	5.428
Bluetooth connection (Claimer side)	0.320
Bluetooth connection (Verifier side)	0.017
Signing message	0.010
Verifying message	0.004

Bluetooth connection's energy consumption will add up with the increase in the number of verifiers per claim.

End-to-End Performance Results:

One claimer and different number of verifiers. Talasila et al. performed this set of experiments to see whether the estimation of latency and energy consumption based on the micro-benchmark results is correct. In these experiments the authors measured the LINK RTT and the total energy consumption at the claimer, while varying the number of verifiers. The results in Figure 8.18 demonstrates that in terms of latency, the estimations are highly accurate. In terms of power, the authors observed a linear increase in the difference between the estimations and the measured values. The difference becomes significant ($>25\%$) for 5 verifiers. This difference could be explained by the extra energy spent by Bluetooth to create and maintain Piconets with a higher number of slaves.

The relevance of these results comes in the context of the two main questions the authors aimed to answer: (1) Can LINK work well in the presence of mobility? and (2) Can LINK run on the phones without quickly exhausting the battery?

If both the claimer and verifiers move at regular walking speed (1.2m/s), then LINK RTT should be less than 8s to have any chance to establish Bluetooth connections before users move out of each other's transmission range (i.e., Bluetooth range is 10m). This bound is the worst-case scenario, and LINK could barely provide a response with just one verifier. Of course, LINK would perform better if not all users move or move in the same direction. A simple solution to avoid this worst-case scenario is to make the claimer aware that she should not walk while submitting claims. This is a reasonable requirement because the claimer needs to interact with the phone to access the LBS anyway. In such a case, the bound on RTT doubles to 16s, and LINK will be able to establish connections with all verifiers (as shown in Figure 8.18). Note that a Piconet is limited to 7 slaves, which limits the maximum RTT.

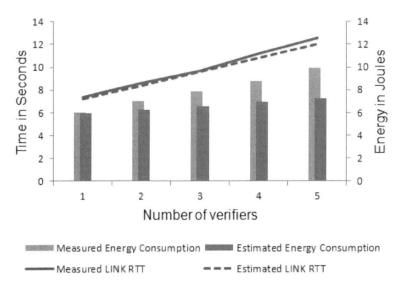

FIGURE 8.18
LINK RTT and total energy consumed by claimer per claim function of the number of verifiers.

TABLE 8.7
Battery life for different WiFi and Bluetooth radio states

	Bluetooth and WiFi off	Bluetooth off and WiFi on	Bluetooth and WiFi on
Battery life	10Days 16Hrs	3Days 15Hrs	2Days 11Hrs

Finally, let us recall that LINK is robust to situations when verifiers run out of the transmission range before the Bluetooth connection is established.

To understand LINK's feasibility from an energy point of view (i.e., to answer the second question posted above), Talasila et al. performed an analysis to see how many claims and verifications a smartphone can execute before it runs out of battery power. The total capacity of the Motorola Droid 2 battery is 18.5KJ. Since LINK requires WiFi and Bluetooth to be on, the authors first measured the effect of these wireless interfaces on the phone lifetime. Table 8.7 shows the results for different interface states (without running any applications). The authors observed that even when both are on all the time, the lifetime is still over 2 days, which is acceptable (most users re-charge their phones at night). The lifetime is even better in reality because Android puts WiFi to sleep when there is no process running that uses WiFi (in the experiments, the authors forced it to be on all the time).

Next, using this result, the authors estimated how many claims and verifications LINK can do with the remaining phone energy:

TABLE 8.8

Average RTT and energy consumption per claim for multi-claimer case vs. single claimer case

	Average RTT (s)	Energy consumed (Joules)
Multi-Claimer case	15.31	9.25
Single-Claimer case	8.60	7.04

*Number of claims a phone can do until battery is exhausted = **2,701** Claims*

*Number of verifications a phone can do until battery is exhausted = **20,458** Verifications*

These numbers demonstrate that a fully charged phone is capable of running LINK in real life with minimal impact on the overall battery consumption.

LINK is designed to work for cellular (over 3G) as well, which uses less battery power compared to WiFi [36]. In the experiments, Talasila et al. used WiFi because the phones did not have cellular service plans. In general, LINK will consume even less energy over the cellular network, but its RTT will increase because 3G has typically higher latency than WiFi.

Simultaneous nearby claimers. The goal in this set of experiments is to evaluate how much the RTT and the energy consumption increase when phones act simultaneously as both claimer and verifier. Talasila et al. used three phones that are in the Bluetooth transmission range of each other while sending concurrent claims. For each claim, the other two phones act as verifiers.

When multiple phones send claims simultaneously, hence perform Bluetooth discovery at the same time, the authors notice many situations when only one or even no verifier was discovered. In total, only in 35% of the tests all the verifiers were discovered. This behavior is mostly due to the well-known Bluetooth inquiry clash problem [139]. An additional problem is the shortened discovery period (5.12s instead of 10.24s). While simultaneous claims in the same transmission range are expected to be rare in practice, the authors decided to provide a solution for this problem nevertheless.

Since LCA is aware of all the devices and their claims, it can successfully predict and prevent the Bluetooth inquiry clash by instructing the claimers on when to start their Bluetooth discovery. Practically, a claimer is delayed (for 3s in the experiments) if another one just started a Bluetooth discovery. The delay setting can be determined more dynamically by LCA at runtime depending on factors such as the number of simultaneous claimers and the number of surrounding verifiers in the region. Furthermore, claimers are allowed to perform complete Bluetooth inquiry scans (10.24s).

Table 8.8 compares the RTT and energy consumption measured when simultaneous claims are performed by three phones, with the values measured for the case with one single claimer and two verifiers. With the new settings, the claimers were able to discover all verifiers. However, as expected, this robustness comes at the cost of higher latency (the energy consumption also increases, but not as much as the latency). The good news is that the authors expected all three claimers to be static according to the guidelines for LINK claimers. Therefore, the increased latency should not impact the successful completion of the protocol.

8.3.3   The SHIELD Scheme for Data Verification in Participatory Sensing Systems

The openness of participatory sensing systems is a double edged-sword: On the one hand, it allows a broad participation, but on the other hand it is vulnerable to pollution attacks, in which malicious users submit faulty data seeking to degrade or influence the output of the sensing system. Traditional cryptographic techniques based on authentication, authorization, and access control can prevent outsiders (i.e., external entities, not part of the sensing campaign) from polluting the sensing task. However, such techniques are not immediately effective against attacks from insiders, which are compromised devices that have the necessary credentials to bypass the authentication mechanism and can pollute the sensed data.

Gidsakis et al. [76] propose SHIELD, a novel data verification framework for participatory sensing systems that combines evidence handling (Dempster-Shafer theory) and data-mining techniques to identify and filter out erroneous (malicious) user contributions.

SHIELD considers four types of entities:

- Users: Each user has a mobile device and collects and reports sensory data to the participatory sensing system; users can also query the system about the results of the sensing tasks).

- Campaign administrators: These are the entities that initiate data collection campaigns.

- Identity and credential management infrastructure: Handles user registration and provides authentication, authorization, and access control services.

- Reporting service (RS): Registered users use this service to send data and query the results of a sensing task.

Compromised devices may try to send faulty data in order to pollute the sensing task and can coordinate their efforts in doing so (i.e., attackers can collude). The area of interest of a sensing task is divided into spatial units, which are homogeneous areas with regard to the sensed phenomenon (e.g., road links; or bus or train lines). In other words, the sensed phenomenon does not exhibit significant spatial variations in a spatial unit. Unlike some other prior work, SHIELD seeks to assess user-contributed information without prior statistical descriptions or models of the data to be collected.

SHIELD can complement any security architecture. SHIELD handles available, contradicting evidence, classifies efficiently incoming reports, and effectively separates and rejects those that are faulty. The system works in three phases. In the first phase (bootstrapping), the user sensing reports for each individual spatial unit are classified as non-faulty or faulty based on the other sensing reports. In the second phase (region merging and training), neighboring spatial units that have similar empirical distributions of user reports are merged into larger regions. Training is performed in these resulting regions. The third phase (classification) uses supervised learning that uses the training data from the previous phases in order to classify subsequent user reports.

Finally, we note that SHIELD assumes an identity infrastructure that is not vulnerable to Sybil attacks [67]. Such protections can be achieved at an additional cost [60, 75].

8.3.4 Data Liveness Validation

One application of crowdsensing is citizen journalism, in which regular citizens use their mobile devices to take photos or record videos documenting various events they are witnessing. In this context, it is important to establish the quality of the videos, in order to prevent pollution from malicious users that provide fake videos.

An important aspect when establishing the authenticity of citizen journalism videos is the problem of video "liveness" verification, which is about determining whether the visual stream uploaded by a user has been captured live on a mobile device, and has not been tampered with by a malicious user attempting to game the system.

Rahman et al. [137] introduced Movee, a motion-sensor-based video liveness verification system. Movee requires that users install an application on their devices, which will record a stream of data from the device's inertial sensor (e.g., accelerometer) in addition to the video stream. This leverages the reality that many mobile devices are equipped with accelerometer sensors. Movee exploits the fact that video frames and accelerometer data captured simultaneously will be correlated, and that such a correlation will be difficult to fabricate and emulate by a malicious user. On the back end, a service provider that receives the video and the accelerometer streams establishes the authenticity of the video by checking if the two streams are consistent.

The system consists of four modules. The Video Motion Analysis module uses video processing techniques on the video stream captured by the camera, and produces a time-dependent motion vector. The Inertial Sensor Motion Analysis module converts the data signal captured from the inertial sensor into another time-dependent motion vector. A third module, the Similarity Computation module, computes the similarity of these two motion vectors, producing a set of features that summarize the nature of the similarity. Finally, the Classification module runs a trained classifier on the features in order to decide whether the two motion vectors are correlated.

Movee was designed to mitigate a wide range of attacks, including:

- Copy-paste attack: The adversary is using a man in the middle software to record video and then outputs the result.

- Projection attack: The adversary points the camera of the mobile device to the projection of a video and then outputs the result.

- Direction sync attack: The adversary infers the dominant motion direction of a video and then uses the device to capture an acceleration sample that encodes the same motion direction.

- Cluster attack: The adversary captures a dataset of videos and associated sensor streams, uses a clustering algorithm to group the videos based on their movement, selects a sensor stream from the cluster that is closest to the target video, and then outputs that sensor stream together with the target video.

- Replay attack: The adversary analyzes the target video and then uses a mobile device to create a sensor stream that emulates the movements observed in the target video.

Through extensive experiments, the authors shows that Movee can efficiently differentiate fraudulent videos from genuine videos. The experiments show that Movee achieves a detection accuracy that ranges between 68% and 93% on a Samsung Admire smartphone, and between 76% and 91% on a Google Glass device.

The Movee system has several limitations. First, it is not transparent to its users: The user is required to perform a verification step in the beginning of shooting the video, during which the user needs to move the camera of the device for 6 seconds in a certain direction. Second, Movee is vulnerable to a "stitch" attack, in which the attacker creates a fraudulent video by first live recording a genuine video and then pointing the camera to a pre-recorded target video. These limitations are addressed by Rahman et al. [136], who proposed Vamos, a Video Accreditation through Motion Signatures system. Vamos provides liveness verifications for videos of arbitrary length, is resistant to a wide range of attacks, and is completely transparent to the users; in particular, it requires no special user interaction, nor change in user behavior.

To eliminate the initial verification step that was required in Movee, Vamos uses the entire video and acceleration streams for verification purposes. Vamos consists of a three-step process. First, it divides the input sample into equal length chunks. Second, it classifies each chunk as either genuine or fraudulent. Third, it combines the results of the second step with a suite of novel features to produce a final decision for the original sample.

Vamos was validated based on two sets of videos: The first set consists of

150 citizen journalism videos collected from YouTube; the second set consists of 160 free-form videos collected from a user study. The experimental results show that the classification performance depends on the video. This leads the authors to conclude that the success rate of attacks against video liveness depends on the type of motions encoded in the video. The authors propose a general classification of videos captured on mobile devices, based on user motion, camera motion, and distance to subject.

Other work in this space includes InformaCam [15], which provides mechanisms to ensure that the media was captured by a specific device at a certain location and time. A limitation of InformaCam is that it is vulnerable to projection attacks.

8.3.5 Truth Discovery

Users may contribute conflicting and noisy data in a crowdsensing system. When such user-contributed data is aggregated, an important challenge is to discover the true value of the contributed data. Methods such as voting or averaging are not effective when the majority of users report false data. Ideally, the crowdsensing system should be able to differentiate between reliable and unreliable users. However, the degree of reliability of users is not known a priori. Truth discovery methods [104, 105, 170] have been proposed to tackle the challenge of discovering both true information and user reliability. Such methods rely on the principle that reliable users tend to report true information and truth should be reported by many reliable users. The common principle shared by truth discovery mechanisms is that a particular user will have higher weight if the data provided by the user is closer to the aggregated results, and a particular user's data will be counted more in the aggregation procedure if this user has a higher weight.

Many truth discovery methods do not take into account correlations that exist among sensed entities. For example, neighboring road segments may share similar traffic conditions, and geographically close locations may share similar weather conditions.

Meng et al. [116] propose a truth discovery framework for crowdsensing of correlated entities. They consider a crowdsensing system in which users collect information via smartphones about some entities from a correlated set of entities, and provide this information to a central server. They formulate the task of inferring user reliability and truths as an optimization problem, in which the proposed objective function measures the differences between the user-input observations and the unknown truths, and integrates users' reliabilities as unknown weights.

It remains to be seen if truth discovery methods can handle more sophisticated adversaries, who alternate between providing reliable and unreliable data, or who collude with each other to coordinate their contributions and attempt to defeat statistical-based techniques.

8.4   Conclusion

This chapter examined several security-related issues that may affect the quality of the data collected by mobile crowdsensing systems. We first presented general reliability issues associated to sensed data. We then discussed solutions to ensure the reliability and quality of the sensed data, such as the ILR, LINK, and SHIELD schemes. Finally, we examined solutions to ensure data liveness and truth discovery.

9

Privacy Concerns and Solutions

9.1 Introduction

The increase in user participation and the rich data collection associated with it are beneficial for mobile crowdsensing systems. However, sensitive participant information may be revealed in this process, such the daily routines, social context, or location information. This raises privacy concerns, and depending on the severity of the collected sensitive information, participants may refuse to engage in sensing activities — a serious problem for mobile crowdsensing systems. This chapter discusses potential solutions to the privacy issues introduced by mobile crowdsensing platforms.

9.2 Participant Privacy Issues

Many crowdsensing applications collect data from participants, which can then be used to identify those individuals. Such data may include, for example, different tastes and interests of a particular user, as well as location data. It should be noted that opportunistic crowdsensing raises more privacy concerns than participatory crowdsensing. This is because the users of applications using opportunistic crowdsensing are not in direct control of the data they submit. For example, the GPS data automatically collected on smartphones simply tags a user's location with other sensor data, but when this location trace is aggregated in time based on the historical data, then other user information can be inferred.

GPS sensor readings can be utilized to infer private information about individual users, such as the routes they take during their daily commutes, and their home and work locations. However, the daily commute GPS sensor measurements can be shared within a larger community to obtain traffic congestion levels in a given city. Therefore, it is important to preserve the privacy of individuals, but at the same time enable mobile crowdsensing applications.

9.2.1 Anonymization

A popular approach for preserving privacy of the data is anonymization [153]. Anonymization can be used to remove the identifying information collected by crowdsensing applications, but it raises two problems. Firstly, the mere removal of identifying information like names and addresses from the data cannot guarantee anonymity. For example, in cases where the crowdsensing applications collect location data, the anonymization will not be able to prevent the identification of individuals. The reason is that anonymized location data will still show the frequently visited locations of a person, which in turn may lead to the identification of that person. Secondly, in the context of data anonymization, data utility and data privacy are conflicting goals. As a result, the anonymization of data will enhance privacy protection, but decrease data utility.

9.2.2 Encryption

The privacy of users' personal data can be achieved by using encryption techniques [183]. By encrypting data submitted by the users, unauthorized third parties will not be able to use personal data, even if they acquire access to the encrypted data. However, such cryptographic techniques may be compute-intensive, which leads to increased energy consumption, and may not be scalable because they require the generation and maintenance of multiple keys.

9.2.3 Data Perturbation

Data perturbation [34] refers to adding noise to sensor data before it is submitted by the individuals. As a result of the perturbation, the data submitted by individuals will not be identifiable. Nevertheless, such data would enable good operation of crowdsensing applications.

One popular form of data perturbation is the micro-aggregation. The term micro-aggregation refers to replacing a specific field with an aggregate or more general value. The replacement of a ZIP code with the name of a province or a state is a typical example of micro-aggregation. Micro-aggregation can be operationally defined in terms of two steps, namely, partition and aggregation. Partition refers to partitioning the data set into several parts (groups, clusters). The aggregation refers to replacement of each record in a part with the average record.

Finally, we argue that the privacy is very user specific, as each individual has a different perception of privacy. For example, one person may be willing to share his or her location information continuously whereas another may not. Thus, it is desirable to develop privacy techniques that address variation in individual preferences. Furthermore, a generic framework for data perturbation needs to be developed such that privacy and security can be achieved in a generic setting independent of the nature of the data being shared. Hence, the mobile crowdsensing systems have to find a way to protect the data of individuals while at the same time enabling the operation of the sensing applications. We discuss in the following sections solutions proposed to address the privacy concerns of participants.

9.3 Privacy-Preserving Architectures

Although many mobile crowdsensing systems have been proposed and built, most lack an appropriate privacy component. The existing solutions discussed in Chapter 8 introduce new issues. For example, one of the most important problems is that of the quality of the information provided by the MCS system to the end users. The problem is that, in order to protect the privacy of the participants, most privacy-preserving mechanisms modify their real locations, which makes the reported data as if it had been measured from a different location, introducing noise or false information in the system and to the end users. Another important problem is that of the energy consumption. Privacy-preserving mechanisms consume extra energy and users are not willing to use MCS applications if their batteries are drained considerably faster.

9.3.1 Hybrid Privacy-Preserving Mechanism

Vergara-Laurens et al. [167] proposed a hybrid privacy-preserving mechanism for participatory sensing systems that combines anonymization, data obfuscation, and encryption techniques to increase the quality of information and privacy protection without increasing the energy consumption in a significant manner. A novel algorithm is proposed that dynamically changes the cell sizes of the grid of the area of interest according to the variability of the variable of interest being measured and chooses different privacy-preserving mechanisms depending on the size of the cell. In small cells, where users can be identified more easily, the algorithm uses encryption techniques to protect the privacy of the users and increase the quality of the information, as the reported location is the real location. On the other hand, anonymization and data obfuscation techniques are used in bigger cells where the variability of the variable of interest is low and therefore it is more important to protect the real location (privacy) of the user.

9.3.1.1 System and Threat Model

The system architecture proposed by Vergara-Laurens et al. [167] comprises four main entities: 1) *mobile nodes* to perform the sensing task, run the privacy mechanism, and transmit the data to the application server; 2) *data broker* responsible for sending the list of points of interest to mobile nodes, receiving the sensing reports, validating the mobiles' signature, performing the initial data aggregation, and forwarding the anonymized data to the application server; 3) *application server* responsible for performing the inference process and providing the information to the final users; and 4) *points of interest generator server* responsible for dividing the target area into cells and assigning locations to points of interest. This process is crucial in order to guarantee data privacy and a good quality of information.

The threat model considers both internal and external adversaries. Adversaries can eavesdrop on the communication channel, and they can compromise any system component, including the data broker, the application server, or both.

9.3.1.2   Proposed Privacy Mechanism

The proposed privacy mechanism is a combination of the Points of Interest (POI) mechanism [166] and the Double-encryption technique [87]. First, the *POI mechanism* is designed to mask the real location of the participants using a set of pre-defined locations called points of interest. Each Voronoi space defined for each POI is called a cell. Therefore, before reporting to the data broker, the mobile device changes the actual location on the record for the location of the closest POI. The main advantage of the POI mechanism is the fact that there is no need to have a centralized entity knowing the actual locations of the participants. Besides, when the data broker receives the masked data, it computes the average of the reported data from all the participants in the same cell and reports only one record per cell to the application server. Thus, this mechanism anonymizes the masked data, adding an extra layer of privacy to the model. The main problem of this mechanism is the noise introduced in the data as a consequence of changing the real locations of the users.

The *double-encryption scheme* is designed to prevent the association between the participants and their records in the system without modifying the actual data, consequently improving the quality of information provided to the final user. The steps in this scheme are: First, the mobile device encrypts the sensed data (D_i) using the public key of the application server producing $Enc_{AS}(D_i)$. The sensed data (D_i) holds the location reported by the user and the sensor measurements. Second, $Enc_{AS}(D_i)$ is encrypted using the public key of the data broker, and a double-encrypted record $(Enc_{DB}(Enc_{AS}(D_i)))$ is transmitted to the data broker. As the data broker and the application servers are independently managed, the double-encryption mechanism makes it more difficult for an adversary to obtain the necessary data to identify the users. When the data broker receives the reported data, it decrypts those records obtaining $Enc_{AS}(D_i)$ only. Therefore, an adversary hacking the data broker knows who sent the data but has no information about the data itself. Finally, the data broker sends the encrypted record $Enc_{AS}(D_i)$ to the application server, where the record is de-encrypted to obtain the actual data Di.

In this case, only an adversary compromising the application server will know the sense data but cannot link back to the participant who originated it.

Now the important question is when to use the POI or the double-encryption mechanism? To answer this question, a POI determining algorithm is designed that is used by the *Points of Interest generator server* to determine which mechanism needs to be applied in each cell. Before going into the details of the algorithm, the reader needs to understand the inference techniques applied on the sensed data to generate a Map of estimated values (M_{Rt}) that is used as the input to the POI determining algorithm.

Inference techniques are utilized in MCS systems to estimate the variables of interest in those places where data are not available. Kriging [123] is one of the most widely used techniques on spatio-temporal datasets. Kriging is a BLUE (Best Linear Unbiased Estimator) interpolator. That means it is a linear estimator that matches the correct expected value of the population while minimizing the variance of the observations [77]. In inference stage, the application server applies the Kriging technique, using the data reported by the participants. The result of applying Kriging to the reported data is a Map of estimated values (M_{Rt}) for round R_t, for each point in the area of interest. This map presents to the final user the values of the variables of interest over the entire area. Then, based on the current M_{Rt} map, the Points of Interest generator server runs the POI determining algorithm to define the new set of points of interest and Voronoi spaces (or cells) for the next round R_{t+1}.

The *POI determining algorithm* is the most important part of the new hybrid mechanism. The main idea is to calculate the variability of the variable of interest (e.g., pollution, temperature data) after each round and then change the size of the cells according to it. If the measurements present very low variability (measurements are within a small range of variation) inside a cell, the cell will very likely remain of the same size and the mobile device selects the Points of Interest mechanism to obfuscate the data, i.e., the users will send the reports using the location of the POI of their respective cells. On the other hand, if the variable of interest presents high variability (e.g., very different pollution or temperature values in different zones of the same cell), the algorithm will find those zones with different values and create new cells. If these new cells are smaller than a minimum cell size S_{min}, the mobile device selects the double-encryption technique to encrypt the data because otherwise the user might be recognized considerably more easily (the user will be confined to a smaller area). In this manner, when the mechanism obfuscates the data, it protects the privacy of the user more and saves energy, and when it encrypts the data, it provides more accurate information about the variable of interest but it spends more energy.

9.3.2 SPPEAR Architecture

Gisdakis et al. [75] propose a secure and accountable MCS system that preserves user privacy and enables the provision of incentives to the participants.

At the same time, they are after an MCS system that is resilient to abusive users and guarantees privacy protection even against multiple misbehaving MCS entities (servers). They address these seemingly contradicting requirements with the SPPEAR architecture — Security & Privacy-Preserving Architecture for Participatory-Sensing Applications.

9.3.2.1   System Model

The proposed SPPEAR architecture consists of the following main components: 1) *Task Service Providers* who initiate data-collection campaigns, defining the scope and the domain of the sensing tasks (e.g., estimation of traffic congestion from 8 to 10am). 2) *Users* who carry mobile devices (e.g., smartphones, tablets, smart vehicles) equipped with embedded sensors. Mobile devices collect sensory data and report them to the MCS infrastructure. Additionally, involved participants can also query the results of a sensing task. 3) *Group Manager (GM)*, which is responsible for the registration of user devices, issuing anonymous credentials to them. Furthermore, the Group Manager authorizes the participation of devices in various tasks in an oblivious manner, using authorization tokens. 4) *Identity Provider (IdP)*, which offers identity and credential management services (e.g., user authentication and access control, among others) to the MCS system. 5) *Pseudonym Certification Authority (PCA)*, which provides anonymized ephemeral credentials, termed pseudonyms, to the devices; they are used to cryptographically protect (i.e., ensure the integrity and the authenticity of) the submitted samples, or to authenticate devices querying the results of the sensing task. To achieve unlinkability, devices obtain multiple pseudonyms from the PCA. 6) *Sample Aggregation Service (SAS)*, which is responsible for storing and processing the collected data submitted by the user devices. For each authentic submitted sample, the SAS issues a receipt to the device, which later submits it to claim credits for the sensing task. The SAS exposes interfaces that enable any authenticated and authorized user to query for the results of sensing tasks/campaigns. 7) *Resolution Authority (RA)*, which is responsible for the revocation of the anonymity of offending devices (e.g., devices that disrupt the system or pollute the data collection process).

The adversaries considered in this system are three types: external, internal adversaries, and misbehaving infrastructure components.

External adversaries: External adversaries are entities without an association to the MCS system, and therefore, they have limited disruptive capabilities. They can eavesdrop communications (to gather information on task participation and user activities). They might manipulate the data collection process by submitting unauthorized samples or replaying the ones of benign users. They can also target the availability of the system by launching jamming and D(D)oS attacks. The latter attacks are beyond the system scope and, therefore, the system relies on the network operators (e.g., ISPs) for their mitigation.

Internal adversaries: Internal adversaries can be users or MCS system entities that exhibit malicious behavior. Users, or their compromised devices, might contribute faulty measurements or attempt to impersonate other entities and pose with multiple identities (i.e., acting as a Sybil entity). Moreover, adversarial users could try to exploit the incentive mechanisms in an attempt to increase their utility (e.g., coupons, rewards, quotas, receipts) either without offering the required contributions (i.e., not complying with the requirements of the task) or by double-spending already redeemed quotas.

Misbehaving infrastructure components: The components such as: (a) fully compromised entities that exhibit arbitrary malicious behavior, (b) honest-but-curious entities executing the protocols correctly but curious to learn private user data, and (c) colluding entities, collectively trying to harm user privacy.

9.3.2.2 System Overview

The system process starts at the Task Service (TS), which generates sensing tasks and campaigns. Each task is associated with the number of credits, C, that users shall receive from the TS for their participation, as long as they submit at least n reports to the Sample Aggregation Service (SAS). The (C, n) parameters are included in the task description. Once ready, the TS inform the Group Manager (GM) about the newly generated task. Then, the GM initializes a group signature scheme that allows each participant (P_i) to anonymously authenticate herself with a private key ($gski$). The GM pushes the group public key to the Identity Provider (IdP) responsible for authenticating users.

The GM publishes a list of active tasks that users regularly retrieve in order to select the ones they want to contribute to. The task description can be done with the use of task-specific languages similar to AnonyTL [148]. If a user is willing to participate in a task, she authorizes her device to obtain the group credentials (i.e., $gski$) and an authorization token from the GM. Then, the device initiates the authentication protocol with the IdP and it obtains pseudonyms from the Pseudonym Certification Authority (PCA). With these pseudonyms the device can (anonymously) authenticate the samples it submits to the SAS (and receive a credit receipt for each of them) or get authenticated to query the task results (Section 6.4). Finally, the device presents n receipts to the TS to receive the task credits. In this fashion, SPPEAR enables the provision of incentives in a privacy-preserving manner, which is a catalyst for user participation.

9.3.3 Privacy-Preserving Truth Discovery

Users may provide noisy, conflicting, or faulty data to a crowdsensing system. As described in Chapter 8, truth discovery methods seek to improve the aggregation accuracy by discovering truthful facts from unreliable user information. However, truth discovery approaches fail to take into account user privacy. Sensitive individual user data such as health data or location information may be revealed, which may have serious privacy consequences.

To address this issue, Miao et al. [117] propose a cloud-enabled privacy-preserving truth discovery (PPTD) framework. The framework uses homomorphic encryption, more precisely the threshold variant of the Paillier cryptosystem [58]. In the proposed PPTD framework, each participating user sends to a central server the encrypted summation of distances between his own observation values and the estimated aggregated values. The server then updates users' weights in encrypted form and sends the updated weight to each user. Each user then uses the received encrypted weight to calculate the ciphertexts of weighted data. The server can estimate the final results based on the ciphertexts received from users. Thus, this framework can accurately calculate the final aggregated results while protecting both the privacy of user data and the user weights.

9.3.4 Other Solutions

AnonySense [148] is a general-purpose framework for secure and privacy-preserving tasking and reporting. Reports are submitted through wireless access points, while leveraging Mix Networks [51] to de-associate the submitted data from their origin. However, the way it employs the short group signatures scheme defined in [40], for the cryptographic protection of submitted reports, makes it vulnerable to sybil attacks. Although AnonySense can evict malicious users, filtering out their past and faulty contributions requires the de-anonymization of benign reports; besides being a costly operation, this process violates the anonymity of legitimate participants. Misbehavior detection is a lengthy process that may occur even at the end of the sensing task when all contributions are available (e.g., by detecting outliers).

There are other solutions such as group signature schemes, which can prevent anonymity misuse by limiting the rate of user authentications (and, hence, the samples they submit), to a predefined threshold (k) for a given time interval [43]. Exceeding this is considered misbehavior and results in user de-anonymization and revocation. However, this technique cannot capture other types of misbehavior, i.e., when malicious users/devices pollute the data collection process by submitting (k - 1) faulty samples within a time interval. Such user misbehavior in the system and anonymity abuse can be prevented by leveraging authorization tokens and pseudonyms with non-overlapping validity periods similar to techniques used in the SPPEAR architecture [75].

PRISM [59] focuses on the secure deployment of sensing applications and does not consider privacy. It follows the push model for distributing tasks to nodes: service providers disseminate applications to mobile devices (according to criteria such as their location). This approach enables timely and scalable application deployment, but from a security perspective it harms user privacy since service providers have knowledge of the device locations. To protect the privacy of the parties querying mobile nodes, PEPPeR [65] decouples the process of node discovery from the access control mechanisms used to query these nodes.

PEPSI [60] is a centralized solution that provides privacy to data queriers and, at the same time, prevents unauthorized entities from querying the results of sensing tasks. However, PEPSI does not consider accountability and privacy-preserving incentive mechanisms and it does not ensure privacy against cellular Internet Service Providers (ISPs).

9.4 Privacy-Aware Incentives

The large-scale deployment of mobile crowdsensing applications is mostly hindered by the lack of incentives for users to participate and the concerns about possible privacy leakage. For instance, to monitor the propagation of a new flu, a server will collect information on who has been infected by this flu. However, a patient may not want to provide such information if she is not sure whether the information will be abused by the server.

9.4.1 Privacy-Aware Incentive Schemes

Li and Cao [106] propose two privacy-aware incentive schemes for mobile sensing to promote user participation. These schemes allow each mobile user to earn credits by contributing data without leaking which data was contributed. At the same time, the schemes ensure that dishonest users cannot abuse the system to earn an unlimited amount of credits. The first scheme considers scenarios where a trusted third party (TTP) is available. It relies on the TTP to protect user privacy, and thus has very low computation and storage cost at each mobile user. The second scheme removes the assumption of TTP and applies blind signature and commitment techniques to protect user privacy.

The privacy-aware incentive scheme works as follows: *First*, to achieve the incentive goal that each participant can earn at most c credits from each task, the approach used in the scheme satisfies three conditions:

1. Each participant can accept a task at most once.

2. The participant can submit at most one report for each accepted task.

3. The participant can earn c credits from a report.

To satisfy the first condition, the basic idea is to issue one request token for each task to each participant. The participant consumes the token when it accepts the task. Since it does not have more tokens for the task, it cannot accept the task again. Similarly, to satisfy the second condition, each participant will be given one report token for each task. It consumes the token when it submits a report for the task and thus cannot submit more reports. To satisfy the last condition, when the service provider receives a report, it issues pseudo-credits to the reporting participant, which can be transformed to c credit tokens. The participant will deposit these tokens to its credit account.

Second, to achieve the privacy goals, all tokens are constructed in a privacy-preserving way, such that a request (report) token cannot be linked to a participant and a credit token cannot be linked to the task and report from which the token is earned.

Therefore, the scheme precomputes privacy-preserving tokens for participants, which are used to process future tasks. To ensure that participants will use the tokens appropriately in the smartphones (i.e., they will not abuse the tokens), commitments to the tokens are also precomputed such that each request (report) token is committed to a specific task and each credit token is committed to a specific participant's smartphone.

9.4.2   Other Schemes

Several schemes [148, 60, 55] have been proposed to protect user privacy in mobile crowdsensing, but they do not provide incentives for users to participate. Yang et al. [181] designed incentives based on gaming and auction theories, but they did not consider privacy. Although incentive and privacy have been addressed separately in mobile crowdsensing, it is still a major issue to address them simultaneously. SPPEAR [75] is one such system that preserves user privacy, and enables the provision of incentives to the participants. Privacy-aware incentives designed for other applications (e.g., publish–subscribe systems) [95] cannot be directly applied here since they do not consider the requirements of mobile sensing.

9.5   Location and Context Privacy

Location privacy is an important concern in mobile crowdsensing applications, where users can both contribute valuable information (data reporting) as well as retrieve (location dependent) information (query) regarding their surroundings. Privacy concern arises mainly in the query process, where a user sends location-sensitive queries regarding his surroundings (e.g., "where is the closest pub?"). The location privacy mainly concerns two objectives: hide user locations, and hide user identities, which avoids association of users with their activities (e.g., "who is requesting the nearest pub?").

9.5.1 Solutions Based on K-Anonymity

K-anonymity [153] is an important measure for privacy to prevent the disclosure of personal data. A table satisfies K-anonymity if every record in the table is indistinguishable from at least $K-1$ other records with respect to every set of quasi-identifier attributes. In the context of location privacy, the location attribute can be viewed as a quasi-identifier. K-anonymous location privacy thus implies that the user's location is indistinguishable from at least $K-1$ other users. To achieve K-anonymous location privacy, one common approach is to incorporate a trust server, called the anonymizer, who is responsible for removing the user's ID and selecting an anonymizing spatial region (ASR) containing the user and at least $K-1$ users in the vicinity.

Another purpose of ASR is to reduce the communication cost between the anonymizer and the service provider, and the processing time at the service provider side. This process is also called "cloaking" as it constructs a spatial cloak around the user's actual location. The anonymizer forwards the ASR along with the query to the (untrusted) location-based service (LBS), which processes the query and returns to the anonymizer a set of candidate points of interests (POIs). The anonymizer removes the false hits and forwards the actual result to the user.

In achieving K-anonymous location privacy, it is crucial to devise quality spatial cloaks at the anonymizer and efficient searching algorithms at the LBS. Intuitively, the cloaks produced should be locality-preserving, close to the user location, and small in size, since both the computational complexity of the search algorithms and the number of POIs returned increases with the size of the cloak. Vu et al. [169] propose a mechanism based on locality-sensitive hashing (LSH) to partition user locations into groups, each containing at least K users (called spatial cloaks). The mechanism is shown to preserve both locality and K-anonymity.

9.5.2 Other Solutions

The location privacy solutions broadly fall into two categories: user-side approaches and approaches that require a trusted server. In the first category, users anonymize their location-based queries by adding noise to the location attributes or generating multiple decoys at different locations. One such approach is called SpaceTwist [184]. In SpaceTwist, starting with a location different from the user's actual location, the nearest neighbors are retrieved incrementally until the query is answered correctly. The uncertainty of the user location is roughly the distance from the initial location to the user's actual location. SpaceTwist requires implementation of incremental k-NN query on the server sides. Furthermore, it does not guarantee K-anonymity if the resulting uncertain area contains less than $K-1$ other users.

In the second category, with a trusted anonymizer, more sophisticated spatial cloaking mechanisms can be devised. In Casper [122], the anonymizer maintains the locations of the clients using a pyramid data structure, similar to a quad-tree. Upon reception of a query, the anonymizer first hashes the user's location to the leaf node and then moves up the tree if necessary until enough neighbors are included. Hilbert cloaking [98] uses the Hilbert space filling to map 2-D space into 1-D values. These values are then indexed by an annotated B+-tree, which supports efficient search by value or by rank (i.e., position in the 1-D sorted list). The algorithm partitions the 1-D sorted list into groups of K users. Hilbert cloaks, though achieving K-anonymity, do not always preserve locality, which leads to large cloak size and high server-side complexity. Recognizing that Casper does not provide K-anonymity, Ghinita et al. [73] proposed a framework for implementing reciprocal algorithms using any existing spatial index on the user locations. Once the anonymous set (AS) is determined, the cloak region can be represented by rectangles, disks, or simply the AS itself.

In another solution [102], the authors proposed that the way to protect the location privacy of these mobile users is through the use of cloud-based agents, which obfuscate user location and enforce the sharing practices of their owners. The cloud agents organize themselves in to a quadtree that enables queriers to directly contact the mobile users in the area of interest and, based on their own criteria, select the ones to get sensing data from. The tree is kept in a decentralized manner, stored and maintained by the mobile agents themselves, thus avoiding the bottlenecks and the privacy implications of centralized approaches.

The authors in [72] use balanced trees to enforce k-anonymity in the spatial domain and conceal user locations. The emphasis here is on users issuing location-based queries as in the case, for example, where a user asks for all hospitals close to its current location. As these queries may compromise user privacy, the use of trees is necessitated by the need to partition the geometric space and answer quickly queries about the location of other users in order to guarantee k-anonymity. In the solution [102], however, the users want to hide their location from a querier who is asking for data provided by the users. Through an appropriate decomposition of the space that is maintained in a distributed quad tree, as opposed to previous approaches which assume a centralized anonymizer, the users can easily obfuscate their exact location according to their own privacy preferences, without relying on other users as in the k-anonymity approaches.

9.5.3   Application: Trace-Hiding Map Generation

Digital maps (e.g., Google Maps) are generated using satellite imagery and street-level information. They are widely used, but they do not always reflect the most up-to-date map information, especially in areas where cities are often undergoing constructions and renovations. Participatory sensing can be used as an effective mechanism to obtain more accurate and current map information. Typically, if participants contribute their entire trace information (with GPS data) to a central map-generation server, the server is able to generate a high-quality map. Obviously, revealing one's entire location trace may raise serious privacy concerns.

Chen et al. [52] propose PMG, a privacy-preserving map generation scheme, in which participants selectively choose, reshuffle, and upload only a few locations from their traces, instead of their entire trace. The final map is then generated by the server from a set of *unorganized* points. The proposed scheme needs to overcome three main challenges:

1. Given a set of data points provided by individual users, preserve the privacy of each user's trace.

2. Generate a high-quality map using the unorganized collection of reported points.

3. Design a map generation scheme that is robust to noisy data such as GPS errors.

To tackle the first challenge, users employ two techniques. Each user shuffles the points from her trace before providing the data to the server in order to break the temporal correlation between the reported points. Each user also limits the number of points reported from a region during a fixed time window. The degree of privacy leakage is a function of the number of reported locations. The goal is to achieve trace privacy, i.e., the server should not be able to (uniquely or approximately) recover the trace from the reported points.

The second challenge is tackled using *curve reconstruction* techniques from computational geometry. The system can guarantee that when the set of points reported by all participants reaches a certain threshold, the quality of the generated map is assured.

Regarding the third challenge, the task of generating a map is made difficult by several factors. The sensed GPS data has errors, as GPS readings typically have errors of at least 10 meters. Moreover, the sampled locations are sparse due to the fact that participants only provide a limited number of points in order to preserve the privacy of their traces. The proposed scheme addresses these by filtering GPS data to remove all potential unreliable data and by requesting sufficiently dense samples and carefully clustering the reported sample points. A critical component of the map generation procedure is to query the crowd for points that will produce the best map under a set of constraints on individual location privacy exposure. This is an NP-hard problem, which is solved by the PMG system using a simple heuristic that is shown to provide a solution within a constant factor of the optimum.

9.5.4　Context Privacy

Context-aware applications offer personalized services based on the operating conditions of smartphone users and their surrounding environments. Such applications use sensors such as GPS, accelerometer, proximity sensors, and microphone to infer the context of the smartphone user. Many such applications aggressively collect sensing data, but do not offer users clear statements on how the collected data will be used.

Approaches designed to provide location privacy will not protect the users' context privacy due to the dynamics of user behaviors and temporal correlations between contexts. For example, consider a context-aware application that learns that a user follows a typical trajectory, i.e., the user goes to a coffee shop and then goes to a hospital. If the user discloses that she is at the coffee shop (i.e., a non-sensitive context), this may reveal that the user is likely to go to the hospital next (i.e., a sensitive context).

MaskIt [78] protects the context privacy against an adversary whose strategy is fixed and does not change over time. Such an adversarial model only captures offline attacks, in which the attacker analyzes a user's fixed personal information and preferences.

Wang and Zhang [172] consider stronger adversaries, whose strategy adapts in time depending on the user's context. For example, a context-aware application may sell users' sensing data, and unscrupulous advertisers may push context-related ads to users. The adversary is able to obtain the released sensing data at the time when an untrusted application accesses the data. The adversary can only retrieve a limited amount of data due to computational constraints or limited bandwidth. As a result, the adversary can adaptively choose different subsets of sensors to maximize its long-term utility. All of this is captured in modeling a strategic adversary, i.e., a malicious adversary that seeks to minimize users' utility through a series of strategic attacks.

The overall goal of the authors is to find the optimal defense strategy for users to preserve privacy over a series of correlated contexts. As the user and the adversary have opposite objectives, their dynamic interactions can be modeled as a zero-sum game. Moreover, since the context keeps changing over time and both the user and the adversary perform different actions at different times, the zero-sum game is in a stochastic setting.

The authors model the strategic and dynamic competition between a smartphone user and a malicious adversary as a zero-sum stochastic game, where the user preserves context-based service quality and context privacy against strategic adversaries. The user's action is to control the released data granularity of each sensor used by context-aware applications in a long-term defense against the adversary, while the adversary's action is to select which sensing data as the source for attacks. The user's optimal defense strategy is obtained at a Nash Equilibrium point of this zero-sum game.

The efficiency of the algorithm proposed by the authors to find the optimal defense strategy was validated on smartphone context traces from 94 users. The evaluation results can provide some guidance in the design of future context privacy-preserving schemes.

9.6   Anonymous Reputation and Trust

In participatory sensing systems, ensuring the anonymity of participants is at odds with establishing the trustworthiness of the sensed data. Achieving the right balance between these two is a major challenge. Anonymity is desirable because in participatory sensing, the participants share sensed data tagged with spatio-temporal information, which can reveal sensitive personal details about the participants, such as user identity, personal activities, political views, or health status. Data trustworthiness is necessary because participants can easily provide falsified or unreliable data.

ARTSense [173] is a framework that tries to solve the problem of "trust without identity" in participatory sensing systems. ARTSense assumes that participants submit data through an anonymous network, such as Onion Routing and Mix networks. It also assumes that spatial and temporal cloaking are used at the application level in order to prevent immediate identification of participants.

ARTSense follows the general model for participatory sensing systems, and assumes a central server that receives sensing reports from participants, analyzes them, and makes them available to data consumers. The server maintains an anonymous reputation system for participants. The server establishes the trustworthiness of a sensing report based on two types of provenance. One is *user provenance*, which contains a reputation certificate issued by the server. The other is *contextual provenance*, in which the server compares the information in this report (e.g., location and time) to other reports and assigns a trust value to this report. The server then issues a blinded *reputation feedback coupon* to the user. The user unblinds the feedback coupon and redeems it with the server, which leads to an update in the user's reputation.

9.7   Conclusion

In this chapter, we discussed privacy issues associated with mobile crowdsensing and reviewed several solutions to address them. We first discussed the privacy-preserving architectures based on their system model and adversary setup. Subsequently, we presented privacy-aware incentives, which are crucial to maintain enough participation in mobile crowdsensing. Finally, we discussed solutions for location and context-specific privacy, and for anonymous reputation management. All of these provide privacy assurances to the participants of mobile crowdsensing systems.

10

Conclusions and Future Directions

10.1 Conclusions

The three components of the ubiquitous computing vision are computing, wireless communication, and wireless sensing. With the widespread use of smartphones, computing and wireless communication are on their way to achieving ubiquity. Wireless sensing has been a niche technology until very recently. However, this situation is changing. Encouraged by the ever-expanding ecosystem of sensors embedded in mobile devices and context-aware mobile apps, we believe that our society is on the verge of achieving ubiquitous sensing, and mobile crowdsensing is the enabling technology. In the very near future, crowdsensing is expected to enable many new and useful services for society in areas such as transportation, healthcare, and environment protection.

The main outcome of this book will hopefully be to set the course for mass adoption of mobile crowdsensing. Throughout the book, we have systematically explored the state of the art in mobile crowdsensing and have identified its benefits and challenges. The main advantages of this new technology are its cost-effectiveness at scale and its flexibility to tailor sensing accuracy to the needs and budget of clients (people or organizations) collecting the data. The current crowdsensing systems and platforms have already proved these benefits.

10.2 Future Directions

As described in this this book, there are sound solutions for each of the individual challenges faced by mobile crowdsensing: incentives, privacy, security, and resource management at the mobile devices. The only exception is data reliability, for which we need to expand the current solutions based on location and co-location authentication to solutions that can tell whether the data itself is valid. The remaining challenges faced by mobile crowdsensing lie at the intersection of incentives, privacy, security, and resource management. For example, how can we provide participant anonymity and authenticate their location at the same time? Or how can we balance incentives and resource management when multiple clients attempt to collect data from the same set of participants? New protocols and algorithms will need to be designed and implemented to solve these types of questions.

A different type of remaining challenge is the relation between global sensing tasks, as seen by clients, and the individual sensing tasks executed by participants. Issues such as global task specification, global task decomposition, and individual task scheduling and fault tolerance will need to be investigated in order to allow clients to execute more complex global sensing tasks. For example, in an emergency situation, an application may need access to multiple types of sensors, and these types are defined as a function of region and time. Furthermore, the application should be allowed to specify the desired sensing density, sensing accuracy, or fault tolerance.

In the next 5 to 10 years, we expect smart cities to fully incorporate crowdsensing in their infrastructure management systems. In addition, we believe that crowdsensing and the Internet of Things will complement each other in sensing the physical world. Furthermore, public cloud and cloudlets deployed at the edge of the networks are expected to make crowdsensing and the Internet of Things more effective and efficient. Substantial research and development is needed to take advantage of this infrastructure, which incorporates mobility, sensing, and the cloud. Nevertheless, we are optimistic and fully expect to see complex applications and services running over this type of infrastructure during the next decade.

Bibliography

[1] 99designs crowdsourcing platform. `https://99designs.com`. Retrieved July 4th, 2016.

[2] Alien vs. Mobile User Game website. `http://web.njit.edu/~mt57/avmgame`. Retrieved July 4th, 2016.

[3] Amazon Mechanical Turk. `http://www.mturk.com`. Retrieved July 4th, 2016.

[4] CNN iReport. `http://ireport.cnn.com/`. Retrieved Apr. 30th, 2015.

[5] Crowdflower crowdsourcing platform. `https://www.crowdflower.com`. Retrieved July 4th, 2016.

[6] Dell's IdeaStorm Project. `http://www.ideastorm.com`. Retrieved July 4th, 2016.

[7] FEMA Mobile App. `http://www.fema.gov/mobile-app`. Retrieved Mar. 6th, 2016.

[8] Galaxy Zoo Project. `https://www.galaxyzoo.org`. Retrieved July 4th, 2016.

[9] Garmin, Edge 305. `www.garmin.com/products/edge305/`. Retrieved Nov. 12th, 2014.

[10] Global smartphone shipments forecast. `http://www.statista.com/statistics/263441/global-smartphone-shipments-forecast/`. Retrieved Mar. 13th, 2015.

[11] Google Mobile Ads. `http://www.google.com/ads/mobile/`. Retrieved July 4th, 2016.

[12] Google Play Android App Store. `https://play.google.com/`. Retrieved July 4th, 2016.

[13] Google's Recaptcha Project. `https://www.google.com/recaptcha`. Retrieved July 4th, 2016.

[14] Hadoop Website. `http://hadoop.apache.org/`. Retrieved July 4th, 2016.

[15] InformaCam: Verified Mobile Media. `https://guardianproject.info/apps/informacam/`. Retrieved July 4th, 2016.

[16] Innocentive Crowdsourcing Platform. `https://www.innocentive.com/`. Retrieved July 4th, 2016.

[17] Intel Labs, The Mobile Phone that Breathes. `http://scitech.blogs.cnn.com/2010/04/22/the-mobilephone-that-breathes/`. Retrieved Nov. 12th, 2014.

[18] Kaggle crowdsourcing platform. `https://www.kaggle.com`. Retrieved July 4th, 2016.

[19] McSense Android Smartphone Application. `https://play.google.com/store/apps/details?id=com.mcsense.app`. Retrieved July 4th, 2016.

[20] Memcached Website. `http://memcached.org/`. Retrieved July 4th, 2016.

[21] MIT News. `http://web.mit.edu/newsoffice/2009/blood-pressure-tt0408.html`. Retrieved Nov. 12th, 2014.

[22] MobAds. `http://www.mobads.com/`. Retrieved July 4th, 2016.

[23] Mobile Millennium Project. `http://traffic.berkeley.edu/`. Retrieved July 4th, 2016.

[24] NASA's Clickworkers Project. `http://www.nasaclickworkers.com`. Retrieved July 4th, 2016.

[25] Photo journalism website. `http://www.flickr.com/groups/photojournalism`. Retrieved Nov. 12th, 2014.

[26] Sensordrone: The 6th Sense of Your Smartphone. `http://www.sensorcon.com/sensordrone`. Retrieved July 4th, 2016.

[27] Sharp Pantone 5. `http://mb.softbank.jp/en/products/sharp/107sh.html/`. Retrieved Mar. 13th, 2015.

[28] Threadless crowdsourcing platform. `https://www.threadless.com`. Retrieved July 4th, 2016.

[29] Topcoder crowdsourcing platform. `https://www.topcoder.com`. Retrieved July 4th, 2016.

[30] Trusted Platform Module. `http://www.trustedcomputinggroup.org/developers/trusted_platform_module`. Retrieved Nov. 12th, 2014.

[31] Twitter. `http://twitter.com/`. Retrieved July 4th, 2016.

[32] Waze Traffic App. `https://www.waze.com/`. Retrieved Mar. 6th, 2016.

[33] Tarek Abdelzaher, Yaw Anokwa, Peter Boda, Jeff Burke, Deborah Estrin, Leonidas Guibas, Aman Kansal, Samuel Madden, and Jim Reich. Mobiscopes for human spaces. *Pervasive Computing, IEEE*, 6(2):20–29, 2007.

[34] Rakesh Agrawal and Ramakrishnan Srikant. Privacy-preserving data mining. In *ACM Sigmod Record*, volume 29, pages 439–450. ACM, 2000.

[35] I.F. Akyildiz, W. Su, Y. Sankarasubramaniam, and E. Cayirci. Wireless sensor networks: a survey. *Computer networks*, 38(4):393–422, 2002.

[36] A. Anand, C. Manikopoulos, Q. Jones, and C. Borcea. A quantitative analysis of power consumption for location-aware applications on smart phones. In *IEEE International Symposium on Industrial Electronics (ISIE'07)*, pages 1986–1991. IEEE, 2007.

[37] Martin Azizyan, Ionut Constandache, and Romit Roy Choudhury. SurroundSense: mobile phone localization via ambience fingerprinting. In *Proceedings of the 15th annual international conference on Mobile computing and networking*, pages 261–272. ACM, 2009.

[38] Xuan Bao and Romit Roy Choudhury. MoVi: mobile phone based video highlights via collaborative sensing. In *Proceedings of the 8th international conference on Mobile systems, applications, and services*, pages 357–370. ACM, 2010.

[39] M. Bishop. The Total Economic Impact Of InnoCentive Challenges. Technical report, Forrester Consulting, 2009.

[40] Dan Boneh, Xavier Boyen, and Hovav Shacham. Short group signatures. In *Advances in Cryptology–CRYPTO 2004*, pages 41–55. Springer, 2004.

[41] Daren C. Brabham. Crowdsourcing as a Model for Problem Solving – An Introduction and Cases. *Convergence: the international journal of research into new media technologies*, 14(1):75–90, 2008.

[42] Z. Butler and D. Rus. Event-based motion control for mobile-sensor networks. In *Workshop, IEEE Press*, pages 139–140. Citeseer, 1998.

[43] Jan Camenisch, Susan Hohenberger, Markulf Kohlweiss, Anna Lysyanskaya, and Mira Meyerovich. How to win the clonewars: efficient periodic n-times anonymous authentication. In *Proceedings of the 13th ACM conference on Computer and communications security*, pages 201–210. ACM, 2006.

[44] Andrew T. Campbell, Shane B. Eisenman, Nicholas D. Lane, Emiliano Miluzzo, and Ronald A. Peterson. People-centric urban sensing. In *Proceedings of the 2nd annual international workshop on Wireless internet*, page 18. ACM, 2006.

[45] S. Čapkun, M. Čagalj, and M. Srivastava. Secure localization with hidden and mobile base stations. In *in Proceedings of IEEE INFOCOM*. Citeseer, 2006.

[46] S. Capkun and J.P. Hubaux. Secure positioning of wireless devices with application to sensor networks. In *INFOCOM'05. 24th Annual Joint Conference of the IEEE Computer and Communications Societies. Proceedings IEEE*, volume 3, pages 1917–1928. IEEE, 2005.

[47] S. Capkun and J.P. Hubaux. Secure positioning in wireless networks. *IEEE Journal on Selected Areas in Communications*, 24(2):221–232, 2006.

[48] G. Cardone, A. Corradi, L. Foschini, and R. Ianniello. ParticipAct: a Large-Scale Crowdsensing Platform. *Emerging Topics in Computing, IEEE Transactions on*, PP(99):1–1, 2015.

[49] Giuseppe Cardone, Andrea Cirri, Antonio Corradi, and Luca Foschini. The ParticipAct mobile crowd sensing living lab: The testbed for smart cities. *Communications Magazine, IEEE*, 52(10):78–85, 2014.

[50] Giuseppe Cardone, Luca Foschini, Cristian Borcea, Paolo Bellavista, Antonio Corradi, Manoop Talasila, and Reza Curtmola. Fostering ParticipAction in Smart Cities: a Geo-Social CrowdSensing Platform. *IEEE Communications Magazine*, 51(6), 2013.

[51] David L Chaum. Untraceable electronic mail, return addresses, and digital pseudonyms. *Communications of the ACM*, 24(2):84–90, 1981.

[52] Xi Chen, Xiaopei Wu, Xiang-Yang Li, Yuan He, and Yunhao Liu. Privacy-preserving high-quality map generation with participatory sensing. In *Proc. of INFOCOM '14*, pages 2310–2318, 2014.

[53] Man Hon Cheung, Richard Southwell, Fen Hou, and Jianwei Huang. Distributed Time-Sensitive Task Selection in Mobile Crowdsensing. *arXiv preprint arXiv:1503.06007*, 2015.

[54] Tanzeem Choudhury, Sunny Consolvo, Beverly Harrison, Jeffrey Hightower, Anthony LaMarca, Louis LeGrand, Ali Rahimi, Adam Rea, G. Bordello, and Bruce Hemingway. The mobile sensing platform: An embedded activity recognition system. *Pervasive Computing, IEEE*, 7(2):32–41, 2008.

[55] Delphine Christin, Christian Roßkopf, Matthias Hollick, Leonardo A. Martucci, and Salil S. Kanhere. Incognisense: An anonymity-preserving reputation framework for participatory sensing applications. *Pervasive and Mobile Computing*, 9(3):353–371, 2013.

[56] X. Chu, X. Chen, K. Zhao, and J. Liu. Reputation and trust management in heterogeneous peer-to-peer networks. *Springer Telecommunication Systems*, 44(3-4):191–203, Aug. 2010.

[57] Bram Cohen. Incentives build robustness in BitTorrent. In *Workshop on Economics of Peer-to-Peer systems*, volume 6, pages 68–72, 2003.

[58] Ivan Damgård and Mats Jurik. A Generalisation, a Simplification and Some Applications of Paillier's Probabilistic Public-Key System. In *Proceedings of the 4th International Workshop on Practice and Theory in Public Key Cryptography: Public Key Cryptography*, PKC '01, pages 119–136. Springer-Verlag, 2001.

[59] Tathagata Das, Prashanth Mohan, Venkata N Padmanabhan, Ramachandran Ramjee, and Asankhaya Sharma. PRISM: platform for remote sensing using smartphones. In *Proceedings of the 8th international conference on Mobile systems, applications, and services*, pages 63–76. ACM, 2010.

[60] Emiliano De Cristofaro and Claudio Soriente. Extended capabilities for a privacy-enhanced participatory sensing infrastructure (PEPSI). *Information Forensics and Security, IEEE Transactions on*, 8(12):2021–2033, 2013.

[61] Lilian de Greef, Mayank Goel, Min Joon Seo, Eric C. Larson, James W. Stout, James A. Taylor, and Shwetak N. Patel. Bilicam: using mobile phones to monitor newborn jaundice. In *Proceedings of the 2014 ACM International Joint Conference on Pervasive and Ubiquitous Computing, UbiComp'14*, pages 331–342. ACM, 2014.

[62] Murat Demirbas, Murat Ali Bayir, Cuneyt Gurcan Akcora, Yavuz Selim Yilmaz, and Hakan Ferhatosmanoglu. Crowd-sourced sensing and collaboration using twitter. In *2010 IEEE International Symposium on World of Wireless Mobile and Multimedia Networks (WoWMoM)*, pages 1–9. IEEE, 2010.

[63] Linda Deng and Landon P. Cox. LiveCompare: grocery bargain hunting through participatory sensing. In *Proceedings of the 10th workshop on Mobile Computing Systems and Applications*, page 4. ACM, 2009.

[64] Y. Desmedt. Major security problems with the unforgeable(feige)-fiat-shamir proofs of identity and how to overcome them. In *SecuriCom (1988)*, volume 88, pages 15–17, 1988.

[65] Tassos Dimitriou, Ioannis Krontiris, and Ahmad Sabouri. PEPPeR: A Queriers Privacy Enhancing Protocol for PaRticipatory Sensing. In *Security and Privacy in Mobile Information and Communication Systems*, pages 93–106. Springer, 2012.

[66] Yi Fei Dong, Salil Kanhere, Chun Tung Chou, and Ren Ping Liu. Automatic image capturing and processing for PetrolWatch. In *2011 17th IEEE International Conference on Networks (ICON)*, pages 236–240. IEEE, 2011.

[67] J. Douceur. The Sybil Attack. In *Proc. of IPTPS '01*, pages 251–260, 2002.

[68] J.S. Downs, M.B. Holbrook, S. Sheng, and L.F. Cranor. Are Your Participants Gaming the System?: Screening Mechanical Turk Workers. In *Proceedings of the 28th international conference on human factors in computing systems (CHI'10)*, pages 2399–2402. ACM, 2010.

[69] Shane B. Eisenman, Emiliano Miluzzo, Nicholas D. Lane, Ronald A. Peterson, Gahng-Seop Ahn, and Andrew T. Campbell. BikeNet: A Mobile Sensing System for Cyclist Experience Mapping. *ACM Transactions on Sensor Networks (TOSN)*, 6(1):6, 2009.

[70] Jakob Eriksson, Lewis Girod, Bret Hull, Ryan Newton, Samuel Madden, and Hari Balakrishnan. The pothole patrol: using a mobile sensor network for road surface monitoring. In *Proceedings of the 6th international conference on Mobile systems, applications, and services*, pages 29–39. ACM, 2008.

[71] Ruipeng Gao, Mingmin Zhao, Tao Ye, Fan Ye, Yizhou Wang, Kaigui Bian, Tao Wang, and Xiaoming Li. Jigsaw: Indoor floor plan reconstruction via mobile crowdsensing. In *Proceedings of the 20th annual international conference on Mobile computing and networking*, pages 249–260. ACM, 2014.

[72] Gabriel Ghinita, Panos Kalnis, and Spiros Skiadopoulos. Prive: anonymous location-based queries in distributed mobile systems. In *Proceedings of the 16th international conference on World Wide Web*, pages 371–380. ACM, 2007.

[73] Gabriel Ghinita, Keliang Zhao, Dimitris Papadias, and Panos Kalnis. A reciprocal framework for spatial k-anonymity. *Information Systems*, 35(3):299–314, 2010.

[74] Dan Gillmor. *We the media: Grassroots journalism by the people, for the people*. O'Reilly Media, Inc., 2006.

[75] Stylianos Gisdakis, Thanassis Giannetsos, and Panos Papadimitratos. SPPEAR: security & privacy-preserving architecture for participatory-sensing applications. In *Proceedings of the 2014 ACM conference on Security and privacy in wireless & mobile networks*, pages 39–50. ACM, 2014.

[76] Stylianos Gisdakis, Thanassis Giannetsos, and Panos Papadimitratos. Shield: A data verification framework for participatory sensing systems. In *Proceedings of the 8th ACM Conference on Security & Privacy in Wireless and Mobile Networks (WiSec '15)*, pages 16:1–16:12. ACM, 2015.

[77] Pierre Goovaerts. *Geostatistics for natural resources evaluation*. Oxford University Press, 1997.

[78] Michaela Götz, Suman Nath, and Johannes Gehrke. Maskit: Privately releasing user context streams for personalized mobile applications. In *Proceedings of the 2012 ACM SIGMOD International Conference on Management of Data*, SIGMOD '12, pages 289–300. ACM, 2012.

[79] Bin Guo, Zhiwen Yu, Xingshe Zhou, and Daqing Zhang. From participatory sensing to mobile crowd sensing. In *Pervasive Computing and Communications Workshops (PERCOM Workshops), 2014 IEEE International Conference on*, pages 593–598. IEEE, 2014.

[80] Ido Guy. Crowdsourcing in the enterprise. In *Proceedings of the 1st international workshop on Multimodal crowd sensing*, pages 1–2. ACM, 2012.

[81] Kyungsik Han, Eric A. Graham, Dylan Vassallo, and Deborah Estrin. Enhancing Motivation in a Mobile Participatory Sensing Project through Gaming. In *Proceedings of 2011 IEEE 3rd international conference on Social Computing (SocialCom'11)*, pages 1443–1448, 2011.

[82] T. He, C. Huang, B.M. Blum, J.A. Stankovic, and T. Abdelzaher. Range-free localization schemes for large scale sensor networks. In *Proceedings of the 9th annual international conference on Mobile computing and networking*, page 95. ACM, 2003.

[83] T. He, S. Krishnamurthy, L. Luo, T. Yan, L. Gu, R. Stoleru, G. Zhou, Q. Cao, P. Vicaire, and J.A. Stankovic. VigilNet: An integrated sensor

network system for energy-efficient surveillance. *ACM Transactions on Sensor Networks (TOSN)*, 2(1):38, 2006.

[84] Gregory W. Heath, Ross C. Brownson, Judy Kruger, Rebecca Miles, Kenneth E. Powell, Leigh T. Ramsey, and Task Force on Community Preventive Services. The effectiveness of urban design and land use and transport policies and practices to increase physical activity: a systematic review. *Journal of Physical Activity & Health*, 3:S55, 2006.

[85] J.C. Herrera, D.B. Work, R. Herring, X.J. Ban, Q. Jacobson, and A.M. Bayen. Evaluation of traffic data obtained via GPS-enabled mobile phones: The Mobile Century field experiment. *Transportation Research Part C: Emerging Technologies*, 18(4):568–583, 2010.

[86] Ryan Herring, Aude Hofleitner, Dan Work, Olli-Pekka Tossavainen, and Alexandre M. Bayen. Mobile millennium-participatory traffic estimation using mobile phones. In *CPS Forum, Cyber-Physical Systems Week 2009*, 2009.

[87] Baik Hoh, Marco Gruteser, Ryan Herring, Jeff Ban, Daniel Work, Juan-Carlos Herrera, Alexandre M. Bayen, Murali Annavaram, and Quinn Jacobson. Virtual trip lines for distributed privacy-preserving traffic monitoring. In *Proceedings of the 6th international conference on Mobile systems, applications, and services*, pages 15–28. ACM, 2008.

[88] Richard Honicky, Eric A. Brewer, Eric Paulos, and Richard White. N-smarts: networked suite of mobile atmospheric real-time sensors. In *Proceedings of the second ACM SIGCOMM workshop on Networked systems for developing regions*, pages 25–30. ACM, 2008.

[89] J. Howe. The Rise of Crowdsourcing. *Wired*, 14(6), 2006.

[90] Wei-jen Hsu, Thrasyvoulos Spyropoulos, Konstantinos Psounis, and Ahmed Helmy. Modeling time-variant user mobility in wireless mobile networks. In *INFOCOM 2007. 26th IEEE International Conference on Computer Communications. IEEE*, pages 758–766. IEEE, 2007.

[91] L. Hu and D. Evans. Localization for mobile sensor networks. In *Proceedings of the 10th annual international conference on Mobile computing and networking*, pages 45–57. ACM New York, NY, USA, 2004.

[92] Xiping Hu, Terry Chu, Henry Chan, and Victor Leung. Vita: A crowdsensing-oriented mobile cyber-physical system. *Emerging Topics in Computing, IEEE Transactions on*, 1(1):148–165, 2013.

[93] Bret Hull, Vladimir Bychkovsky, Yang Zhang, Kevin Chen, Michel Goraczko, Allen Miu, Eugene Shih, Hari Balakrishnan, and Samuel

Madden. Cartel: a distributed mobile sensor computing system. In *Proceedings of the 4th international conference on Embedded networked sensor systems*, pages 125–138. ACM, 2006.

[94] T.E. Humphreys, B.M. Ledvina, M.L. Psiaki, B.W. OHanlon, and P.M. Kintner, Jr. Assessing the spoofing threat: Development of a portable GPS civilian spoofer. In *Proceedings of the ION GNSS International Technical Meeting of the Satellite Division*, 2008.

[95] Mihaela Ion, Giovanni Russello, and Bruno Crispo. Supporting publication and subscription confidentiality in pub/sub networks. In *Security and Privacy in Communication Networks*, pages 272–289. Springer, 2010.

[96] Haiming Jin, Lu Su, Danyang Chen, Klara Nahrstedt, and Jinhui Xu. Quality of information aware incentive mechanisms for mobile crowd sensing systems. In *Proceedings of the 16th ACM International Symposium on Mobile Ad Hoc Networking and Computing*, pages 167–176. ACM, 2015.

[97] Junghyun Jun, Yu Gu, Long Cheng, Banghui Lu, Jun Sun, Ting Zhu, and Jianwei Niu. Social-Loc: Improving indoor localization with social sensing. In *Proceedings of the 11th ACM Conference on Embedded Networked Sensor Systems*, page 14. ACM, 2013.

[98] Panos Kalnis, Gabriel Ghinita, Kyriakos Mouratidis, and Dimitris Papadias. Preventing location-based identity inference in anonymous spatial queries. *Knowledge and Data Engineering, IEEE Transactions on*, 19(12):1719–1733, 2007.

[99] Eiman Kanjo, Steve Benford, Mark Paxton, Alan Chamberlain, Danae Stanton Fraser, Dawn Woodgate, David Crellin, and Adrain Woolard. Mobgeosen: facilitating personal geosensor data collection and visualization using mobile phones. *Personal and Ubiquitous Computing*, 12(8):599–607, 2008.

[100] Aman Kansal, Michel Goraczko, and Feng Zhao. Building a sensor network of mobile phones. In *Proceedings of the 6th international conference on Information processing in sensor networks*, pages 547–548. ACM, 2007.

[101] Aniket Kittur, Boris Smus, Susheel Khamkar, and Robert E. Kraut. Crowdforge: Crowdsourcing complex work. In *Proceedings of the 24th Annual ACM Symposium on User Interface Software and Technology*, UIST '11, pages 43–52. ACM, 2011.

[102] Ioannis Krontiris and Tassos Dimitriou. Privacy-respecting discovery of data providers in crowd-sensing applications. In *Distributed*

Computing in Sensor Systems (DCOSS), 2013 IEEE International Conference on, pages 249–257. IEEE, 2013.

[103] K. Lakhani, K. Boudreau, P. Loh, L. Backstrom, C. Baldwin, E. Lonstein, M. Lydon, A. MacCormack, R. Arnaout, and E. Guinan. Prize-based contests can provide solutions to computational biology problems. *Nature Biotechnolog*, 31(2):108–111, 2013.

[104] Qi Li, Yaliang Li, Jing Gao, Lu Su, Bo Zhao, Murat Demirbas, Wei Fan, and Jiawei Han. A confidence-aware approach for truth discovery on long-tail data. *Proc. VLDB Endow.*, 8(4):425–436, December 2014.

[105] Qi Li, Yaliang Li, Jing Gao, Bo Zhao, Wei Fan, and Jiawei Han. Resolving conflicts in heterogeneous data by truth discovery and source reliability estimation. In *Proceedings of the 2014 ACM SIGMOD International Conference on Management of Data*, SIGMOD '14, pages 1187–1198. ACM, 2014.

[106] Qinghua Li and Guohong Cao. Providing privacy-aware incentives for mobile sensing. In *Pervasive Computing and Communications (PerCom), 2013 IEEE International Conference on*, pages 76–84. IEEE, 2013.

[107] Sophia B. Liu, Leysia Palen, Jeannette Sutton, Amanda L. Hughes, and Sarah Vieweg. In search of the bigger picture: The emergent role of on-line photo sharing in times of disaster. In *Proceedings of the Information Systems for Crisis Response and Management Conference (ISCRAM)*, 2008.

[108] Hong Lu, Wei Pan, Nicholas D. Lane, Tanzeem Choudhury, and Andrew T. Campbell. SoundSense: scalable sound sensing for people-centric applications on mobile phones. In *Proceedings of the 7th international conference on Mobile systems, applications, and services*, pages 165–178. ACM, 2009.

[109] Yu Lu, Shili Xiang, Wei Wu, and Huayu Wu. A queue analytics system for taxi service using mobile crowd sensing. In *Proceedings of the 2015 ACM International Joint Conference on Pervasive and Ubiquitous Computing and Proceedings of the 2015 ACM International Symposium on Wearable Computers*, pages 121–124. ACM, 2015.

[110] Nicolas Maisonneuve, Matthias Stevens, Maria E. Niessen, and Luc Steels. NoiseTube: Measuring and mapping noise pollution with mobile phones. *Information Technologies in Environmental Engineering*, pages 215–228, 2009.

[111] Thomas W. Malone, Robert Laubacher, and Chrysanthos Dellarocas. Harnessing crowds: Mapping the genome of collective intelligence. 2009.

[112] K. Mao, L. Capra, M. Harman, and Y. Jia. A survey of the use of crowdsourcing in software engineering. Technical report, University College London, 2015.

[113] Winter Mason and Duncan J. Watts. Financial Incentives and the Performance of Crowds. *SIGKDD Explor. Newsl.*, 11(2):100–108, 2010.

[114] Suhas Mathur, Tong Jin, Nikhil Kasturirangan, Janani Chandrasekaran, Wenzhi Xue, Marco Gruteser, and Wade Trappe. Parknet: drive-by sensing of road-side parking statistics. In *Proceedings of the 8th international conference on Mobile systems, applications, and services*, pages 123–136. ACM, 2010.

[115] Sam Mavandadi, Stoyan Dimitrov, Steve Feng, Frank Yu, Richard Yu, Uzair Sikora, and Aydogan Ozcan. Crowd-sourced BioGames: managing the big data problem for next-generation lab-on-a-chip platforms. *Lab on a Chip*, 12(20):4102–4106, 2012.

[116] Chuishi Meng, Wenjun Jiang, Yaliang Li, Jing Gao, Lu Su, Hu Ding, and Yun Cheng. Truth discovery on crowd sensing of correlated entities. In *Proceedings of the 13th ACM Conference on Embedded Networked Sensor Systems*, SenSys '15, pages 169–182. ACM, 2015.

[117] Chenglin Miao, Wenjun Jiang, Lu Su, Yaliang Li, Suxin Guo, Zhan Qin, Houping Xiao, Jing Gao, and Kui Ren. Cloud-enabled privacy-preserving truth discovery in crowd sensing systems. In *Proceedings of the 13th ACM Conference on Embedded Networked Sensor Systems*, SenSys '15, pages 183–196. ACM, 2015.

[118] Emiliano Miluzzo, Cory T. Cornelius, Ashwin Ramaswamy, Tanzeem Choudhury, Zhigang Liu, and Andrew T. Campbell. Darwin phones: the evolution of sensing and inference on mobile phones. In *Proceedings of the 8th international conference on Mobile systems, applications, and services*, pages 5–20. ACM, 2010.

[119] Emiliano Miluzzo, Nicholas D. Lane, Kristóf Fodor, Ronald Peterson, Hong Lu, Mirco Musolesi, Shane B. Eisenman, Xiao Zheng, and Andrew T. Campbell. Sensing meets mobile social networks: the design, implementation and evaluation of the CenceMe application. In *Proceedings of the 6th ACM conference on Embedded network sensor systems*, pages 337–350. ACM, 2008.

[120] Prashanth Mohan, Venkata N. Padmanabhan, and Ramachandran Ramjee. Nericell: rich monitoring of road and traffic conditions using mobile smartphones. In *Proceedings of the 6th ACM conference on Embedded network sensor systems*, pages 323–336. ACM, 2008.

[121] Prashanth Mohan, Venkata N. Padmanabhan, and Ramachandran Ramjee. TrafficSense: Rich Monitoring of Road and Traffic Conditions using Mobile Smartphones. Technical Report MSR-TR-2008-59, Microsoft Research, April 2008.

[122] Mohamed F. Mokbel, Chi-Yin Chow, and Walid G. Aref. The new casper: A privacy-aware location-based database server. In *IEEE 23rd International Conference on Data Engineering, ICDE*, pages 1499–1500. IEEE, 2007.

[123] R. Moore. Geostatistics in hydrology: Kriging interpolation. *Mathematics Department, Macquarie University, Sydney., Tech. Rep*, 1999.

[124] Min Mun, Sasank Reddy, Katie Shilton, Nathan Yau, Jeff Burke, Deborah Estrin, Mark Hansen, Eric Howard, Ruth West, and Péter Boda. PEIR, the personal environmental impact report, as a platform for participatory sensing systems research. In *Proceedings of the 7th international conference on Mobile systems, applications, and services*, pages 55–68. ACM, 2009.

[125] Tadao Murata. Petri nets: Properties, analysis and applications. *Proceedings of the IEEE*, 77(4):541–580, 1989.

[126] Sarfraz Nawaz, Christos Efstratiou, and Cecilia Mascolo. ParkSense: A smartphone based sensing system for on-street parking. In *Proceedings of the 19th annual international conference on Mobile computing & networking*, pages 75–86. ACM, 2013.

[127] J. Newsome, E. Shi, D. Song, and A. Perrig. The Sybil attack in sensor networks: analysis & defenses. In *Proc. of IPSN '04*, pages 259–268, 2004.

[128] Claudio E. Palazzi, Gustavo Marfia, and Marco Roccetti. Combining web squared and serious games for crossroad accessibility. In *Serious Games and Applications for Health (SeGAH), 2011 IEEE 1st International Conference on*, pages 1–4. IEEE, 2011.

[129] J. Pan, I. Sandu Popa, K. Zeitouni, and C. Borcea. Proactive vehicular traffic re-routing for lower travel time. *Vehicular Technology, IEEE Transactions on*, 62(8):3551–3568, 2013.

[130] Dan Peng, Fan Wu, and Guihai Chen. Pay as how well you do: A quality based incentive mechanism for crowdsensing. In *Proceedings of the 16th ACM International Symposium on Mobile Ad Hoc Networking and Computing*, pages 177–186. ACM, 2015.

[131] B.S. Peterson, R.O. Baldwin, and J.P. Kharoufeh. Bluetooth inquiry time characterization and selection. *IEEE Transactions on Mobile Computing*, pages 1173–1187, 2006.

[132] Galen Pickard, Iyad Rahwan, Wei Pan, Manuel Cebrian, Riley Crane, Anmol Madan, and Alex Pentland. Time Critical Social Mobilization: The DARPA Network Challenge Winning Strategy. Technical Report arXiv:1008.3172v1, MIT, 2010.

[133] C. Piro, C. Shields, and B. N. Levine. Detecting the Sybil attack in mobile ad hoc networks. In *Proc. of SecureComm'06*, 2006.

[134] M.R. Ra, B. Liu, T.F. La Porta, and R. Govindan. Medusa: A programming framework for crowd-sensing applications. In *Proceedings of the 10th international conference on Mobile systems, applications, and services (MobiSys'12)*, pages 337–350. ACM, 2012.

[135] Kiran K Rachuri, Mirco Musolesi, Cecilia Mascolo, Peter J. Rentfrow, Chris Longworth, and Andrius Aucinas. EmotionSense: a mobile phones based adaptive platform for experimental social psychology research. In *Proceedings of the 12th ACM international conference on Ubiquitous computing*, pages 281–290. ACM, 2010.

[136] M. Rahman, M. Azimpourkivi, U. Topkara, and B. Carbunar. Liveness Verifications for Citizen Journalism Videos. In *Proc. of the 8th ACM Conference on Security and Privacy in Wireless and Mobile Networks (WiSec)*, 2015.

[137] M. Rahman, U. Topkara, and B. Carbunar. Movee: Video liveness verification for mobile devices with built-in motion sensors. *IEEE Transactions on Mobile Computing (TMC)*, 2015.

[138] A. Ranganathan, N. Tippenhauer, B. Škorić, D. Singelée, and S. Čapkun. Design and Implementation of a Terrorist Fraud Resilient Distance Bounding System. *Computer Security–ESORICS*, pages 415–432, 2012.

[139] Nishkam Ravi, Peter Stern, Niket Desai, and Liviu Iftode. Accessing Ubiquitous Services Using Smart Phones. In *Proceedings of the Third IEEE International Conference on Pervasive Computing and Communications (PERCOM'05)*, pages 383–393, Los Alamitos, CA, USA, 2005. IEEE Computer Society.

[140] Lenin Ravindranath, Arvind Thiagarajan, Hari Balakrishnan, and Samuel Madden. Code in the air: simplifying sensing and coordination tasks on smartphones. In *Proceedings of the Twelfth Workshop on Mobile Computing Systems & Applications*, page 4. ACM, 2012.

[141] Sasank Reddy, Deborah Estrin, Mark Hansen, and Mani Srivastava. Examining micro-payments for participatory sensing data collections. In *Proceedings of the 12th ACM international conference on Ubiquitous computing*, pages 33–36. ACM, 2010.

[142] Sasank Reddy, Andrew Parker, Josh Hyman, Jeff Burke, Deborah Estrin, and Mark Hansen. Image browsing, processing, and clustering for participatory DietSense prototype. In *Proceedings of the 4th workshop on Embedded networked sensors*, pages 13–17. ACM, 2007.

[143] Sasank Reddy, Katie Shilton, Gleb Denisov, Christian Cenizal, Deborah Estrin, and Mani Srivastava. Biketastic: sensing and mapping for better biking. In *Proceedings of the SIGCHI Conference on Human Factors in Computing Systems*, pages 1817–1820. ACM, 2010.

[144] John Rula, Vishnu Navda, Fabian Bustamante, Ranjita Bhagwan, and Saikat Guha. No "one-size fits all": Towards a Principled Approach for Incentives in Mobile Crowdsourcing. In *Proceedings of the 15th Workshop on Mobile Computing Systems and Applications (HotMobile)*, pages 3:1–3:5, 2014.

[145] N. Sastry, U. Shankar, and D. Wagner. Secure verification of location claims. In *Proc. of the 2nd ACM Workshop on Wireless Security (Wise'03)*, pages 1–10, Sep. 2003.

[146] L. Selavo, A. Wood, Q. Cao, T. Sookoor, H. Liu, A. Srinivasan, Y. Wu, W. Kang, J. Stankovic, D. Young, et al. Luster: wireless sensor network for environmental research. In *Proceedings of the 5th international conference on Embedded networked sensor systems*, page 116. ACM, 2007.

[147] Katie Shilton, Jeff Burke, Deborah Estrin, Ramesh Govindan, Mark Hansen, Jerry Kang, and Min Mun. Designing the personal data stream: Enabling participatory privacy in mobile personal sensing. TPRC, 2009.

[148] Minho Shin, Cory Cornelius, Dan Peebles, Apu Kapadia, David Kotz, and Nikos Triandopoulos. AnonySense: A system for anonymous opportunistic sensing. *Pervasive and Mobile Computing*, 7(1):16–30, 2011.

[149] Daniel P. Siewiorek, Asim Smailagic, Junichi Furukawa, Andreas Krause, Neema Moraveji, Kathryn Reiger, Jeremy Shaffer, and Fei Lung Wong. SenSay: A Context-Aware Mobile Phone. In *International Symposium on Wearable Computers*, volume 3, page 248, 2003.

[150] D. Singelee and B. Preneel. Location verification using secure distance bounding protocols. In *Proc. of the 2nd IEEE International Conference on Mobile Ad-hoc and Sensor Systems (MASS'05)*, pages 834–840, Nov. 2005.

[151] Noah Snavely, Steven M. Seitz, and Richard Szeliski. Photo tourism: exploring photo collections in 3d. In *ACM transactions on graphics (TOG)*, volume 25, pages 835–846. ACM, 2006.

[152] J. Surowiecki. *The Wisdom of Crowds: Why the Many are Smarter Than the Few and how Collective Wisdom Shapes Business, Economies, Societies, and Nations.* Doubleday, 2004.

[153] Latanya Sweeney. k-anonymity: A model for protecting privacy. *International Journal of Uncertainty, Fuzziness and Knowledge-Based Systems*, 10(05):557–570, 2002.

[154] M. Talasila, R. Curtmola, and C. Borcea. LINK: Location verification through Immediate Neighbors Knowledge. In *Proceedings of the 7th International ICST Conference on Mobile and Ubiquitous Systems, (MobiQuitous'10)*, pages 210–223. Springer, 2010.

[155] Manoop Talasila, Reza Curtmola, and Cristian Borcea. ILR: Improving Location Reliability in Mobile Crowd Sensing. *International Journal of Business Data Communications and Networking*, 9(4):65–85, 2013.

[156] Manoop Talasila, Reza Curtmola, and Cristian Borcea. Improving Location Reliability in Crowd Sensed Data with Minimal Efforts. In *WMNC'13: Proceedings of the 6th Joint IFIP/IEEE Wireless and Mobile Networking Conference.* IEEE, 2013.

[157] Manoop Talasila, Reza Curtmola, and Cristian Borcea. Alien vs. Mobile User Game: Fast and Efficient Area Coverage in Crowdsensing. In *Proceedings of the Sixth International Conference on Mobile Computing, Applications and Services (MobiCASE '14).* ICST/IEEE, 2014.

[158] Manoop Talasila, Reza Curtmola, and Cristian Borcea. Collaborative Bluetooth-based Location Authentication on Smart Phones. *in Elsevier Pervasive and Mobile Computing Journal*, 2014.

[159] Manoop Talasila, Reza Curtmola, and Cristian Borcea. Crowdsensing in the Wild with Aliens and Micro-payments. *IEEE Pervasive Computing Magazine*, 2016.

[160] R. Tan, G. Xing, J. Wang, and H.C. So. Collaborative target detection in wireless sensor networks with reactive mobility. *City University of Hong Kong, Tech. Rep*, 2007.

[161] Evangelos Theodoridis, Georgios Mylonas, Veronica Gutierrez Polidura, and Luis Munoz. Large-scale participatory sensing experimentation using smartphones within a Smart City. In *Proceedings of the 11th International Conference on Mobile and Ubiquitous Systems: Computing, Networking and Services*, pages

178–187. ICST (Institute for Computer Sciences, Social-Informatics and Telecommunications Engineering), 2014.

[162] D. Tian and N.D. Georganas. A node scheduling scheme for energy conservation in large wireless sensor networks. *Wireless Communications and Mobile Computing*, 3(2):271–290, 2003.

[163] K. Toyama, R. Logan, and A. Roseway. Geographic location tags on digital images. In *Proceedings of the eleventh ACM international conference on Multimedia*, pages 156–166. ACM, 2003.

[164] K. Tuite, N. Snavely, D. Hsiao, N. Tabing, and Z. Popovic. PhotoCity: Training Experts at Large-scale Image Acquisition Through a Competitive Game. In *Proceedings of the SIGCHI Conference on Human Factors in Computing Systems*, CHI '11, pages 1383–1392. ACM, 2011.

[165] Guliz S. Tuncay, Giacomo Benincasa, and Ahmed Helmy. Participant recruitment and data collection framework for opportunistic sensing: a comparative analysis. In *Proceedings of the 8th ACM MobiCom workshop on Challenged networks*, pages 25–30. ACM, 2013.

[166] Idalides J. Vergara-Laurens, Miguel Labrador, et al. Preserving privacy while reducing power consumption and information loss in lbs and participatory sensing applications. In *GLOBECOM Workshops (GC Wkshps), 2011 IEEE*, pages 1247–1252. IEEE, 2011.

[167] Idalides J. Vergara-Laurens, Diego Mendez, and Miguel A. Labrador. Privacy, quality of information, and energy consumption in participatory sensing systems. In *Pervasive Computing and Communications (PerCom), 2014 IEEE International Conference on*, pages 199–207. IEEE, 2014.

[168] L. von Ahn and L. Dabbish. Labeling Images with a Computer Game. In *Proceedings of the SIGCHI Conference on Human Factors in Computing Systems*, CHI '04, pages 319–326. ACM, 2004.

[169] Khuong Vu, Rong Zheng, and Lie Gao. Efficient algorithms for k-anonymous location privacy in participatory sensing. In *INFOCOM, 2012 Proceedings IEEE*, pages 2399–2407. IEEE, 2012.

[170] Dong Wang, Lance Kaplan, Hieu Le, and Tarek Abdelzaher. On truth discovery in social sensing: A maximum likelihood estimation approach. In *Proceedings of the 11th International Conference on Information Processing in Sensor Networks*, IPSN '12, pages 233–244. ACM, 2012.

[171] G. Wang, G. Cao, and T.F.L. Porta. Movement-assisted sensor deployment. *IEEE Transactions on Mobile Computing*, 5(6):640–652, 2006.

[172] Wei Wang and Qian Zhang. A stochastic game for privacy preserving context sensing on mobile phone. In *2014 IEEE Conference on Computer Communications (INFOCOM)*, pages 2328–2336. IEEE, 2014.

[173] Xinlei Wang, Wei Cheng, P. Mohapatra, and T. Abdelzaher. ARTSense: Anonymous reputation and trust in participatory sensing. In *Proc. of INFOCOM '13*, pages 2517–2525, 2013.

[174] Yi Wang, Wenjie Hu, Yibo Wu, and Guohong Cao. Smartphoto: a resource-aware crowdsourcing approach for image sensing with smartphones. In *Proceedings of the 15th ACM international symposium on Mobile ad hoc networking and computing*, pages 113–122. ACM, 2014.

[175] J. White, C. Thompson, H. Turner, B. Dougherty, and D.C. Schmidt. WreckWatch: automatic traffic accident detection and notification with smartphones. *Mobile Networks and Applications*, 16(3):285–303, 2011.

[176] Haoyi Xiong, Daqing Zhang, Guanling Chen, Leye Wang, and Vincent Gauthier. CrowdTasker: maximizing coverage quality in piggyback crowdsensing under budget constraint. In *Proceedings of the IEEE International Conference on Pervasive Computing and Communications (PerCom15)*.

[177] Liwen Xu, Xiaohong Hao, Nicholas D. Lane, Xin Liu, and Thomas Moscibroda. More with less: lowering user burden in mobile crowdsourcing through compressive sensing. In *Proceedings of the 2015 ACM International Joint Conference on Pervasive and Ubiquitous Computing*, pages 659–670. ACM, 2015.

[178] N. Xu, S. Rangwala, K.K. Chintalapudi, D. Ganesan, A. Broad, R. Govindan, and D. Estrin. A wireless sensor network for structural monitoring. In *Proceedings of the 2nd international conference on Embedded networked sensor systems*, pages 13–24. ACM New York, NY, USA, 2004.

[179] T. Yan, M. Marzilli, R. Holmes, D. Ganesan, and M. Corner. mCrowd: a platform for mobile crowdsourcing. In *Proceedings of the 7th ACM Conference on Embedded Networked Sensor Systems (SenSys'09)*, pages 347–348. ACM, 2009.

[180] Tingxin Yan, Vikas Kumar, and Deepak Ganesan. Crowdsearch: exploiting crowds for accurate real-time image search on mobile phones. In *Proceedings of the 8th international conference on Mobile systems, applications, and services*, pages 77–90. ACM, 2010.

[181] Dejun Yang, Guoliang Xue, Xi Fang, and Jian Tang. Crowdsourcing to smartphones: incentive mechanism design for mobile phone sensing. In *Proceedings of the 18th annual international conference on Mobile computing and networking*, pages 173–184. ACM, 2012.

[182] Jiang Yang, Lada A. Adamic, and Mark S. Ackerman. Crowdsourcing and Knowledge Sharing: Strategic User Behavior on Taskcn. In *Proceedings of the 9th ACM Conference on Electronic Commerce*, EC '08, pages 246–255. ACM, 2008.

[183] Andrew Chi-Chih Yao. Protocols for secure computations. In *FOCS*, volume 82, pages 160–164, 1982.

[184] Man Lung Yiu, Christian S. Jensen, Xuegang Huang, and Hua Lu. Spacetwist: Managing the trade-offs among location privacy, query performance, and query accuracy in mobile services. In *Data Engineering, 2008. ICDE 2008. IEEE 24th International Conference on*, pages 366–375. IEEE, 2008.

[185] Daqing Zhang, Haoyi Xiong, Leye Wang, and Guanling Chen. CrowdRecruiter: selecting participants for piggyback crowdsensing under probabilistic coverage constraint. In *Proceedings of the 2014 ACM International Joint Conference on Pervasive and Ubiquitous Computing*, pages 703–714. ACM, 2014.

[186] L. Zhang, B. Tiwana, Z. Qian, Z. Wang, R.P. Dick, Z.M. Mao, and L. Yang. Accurate online power estimation and automatic battery behavior based power model generation for smartphones. In *Proceedings of the eighth IEEE/ACM/IFIP international conference on Hardware/software codesign and system synthesis (CODES/ISSS'10)*, pages 105–114. ACM, 2010.

[187] Qingwen Zhao, Yanmin Zhu, Hongzi Zhu, Jian Cao, Guangtao Xue, and Bo Li. Fair energy-efficient sensing task allocation in participatory sensing with smartphones. In *INFOCOM, 2014 Proceedings IEEE*, pages 1366–1374. IEEE, 2014.

Index

Printed and bound by CPI Group (UK) Ltd, Croydon, CR0 4YY

28/10/2024

01780264-0003